Urban Horticulture - Necessity of the Future

Edited by Shashank Shekhar Solankey, Shirin Akhtar, Alejandro Isabel Luna Maldonado, Humberto Rodriguez-Fuentes, Juan Antonio Vidales Contreras and Julia Mariana Márquez Reyes

Published in London, United Kingdom

IntechOpen

Supporting open minds since 2005

Urban Horticulture - Necessity of the Future
http://dx.doi.org/10.5772/intechopen.82900
Edited by Shashank Shekhar Solankey, Shirin Akhtar, Alejandro Isabel Luna Maldonado, Humberto Rodriguez-Fuentes, Juan Antonio Vidales Contreras and Julia Mariana Márquez Reyes

Contributors
Alejandro Isabel Luna Maldonado, Humberto Rodríguez-Fuentes, Héctor Flores Breceda, Urbano Luna Maldonado, Julia Mariana Márquez Reyes, Juan Antonio Vidales Contreras, Lawrence Griffing, Krishna Kumar, Juan Carlos Rodríguez Ortiz, Adriana Maria Dos Santos, Marina Paiva Baracuhy, Dermeval Araújo Furtado, Jackson Rômulo De Sousa Leite, Fabiana Terezinha Leal De Morais, Romulo Wilker Neri De Andrade, Ifeoluwapo Amao, Ileana Paladino, Ana Clara Sokolowski, Ines Babnik, Harshata Pal, Anwesha Chatterjee, Sanjit Debnath, Pramod Kumar, Simran Saini, Rodica V. Ghita, Daniela Predoi, Simona Liliana Iconaru, Carmen Laura Cimpeanu, Stefania Mariana Raita, Prack Mc Cormick Barbara, José Enrique Wolski, Rodríguez Hernán y Mauro Navas

First published in London, United Kingdom, 2020 by IntechOpen
IntechOpen is the global imprint of INTECHOPEN LIMITED, registered in England and Wales, registration number: 11086078, 7th floor, 10 Lower Thames Street, London, EC3R 6AF, United Kingdom
Printed in Croatia

British Library Cataloguing-in-Publication Data
A catalogue record for this book is available from the British Library

Additional hard and PDF copies can be obtained from orders@intechopen.com

Urban Horticulture - Necessity of the Future
Edited by Shashank Shekhar Solankey, Shirin Akhtar, Alejandro Isabel Luna Maldonado, Humberto Rodriguez-Fuentes, Juan Antonio Vidales Contreras and Julia Mariana Márquez Reyes
p. cm.
Print ISBN 978-1-83880-512-8
Online ISBN 978-1-83880-513-5
eBook (PDF) ISBN 978-1-83880-631-6

For EU product safety concerns:
IN TECH d.o.o., Prolaz Marije Krucifikse Kozulić 3, 51000 Rijeka, Croatia, info@intechopen.com or visit our website at intechopen.com.

Meet the editors

Dr. Shashank Shekhar Solankey is presently working as Assistant Professor–cum–Jr. Scientist (Horticulture: Vegetable Science) at Bihar Agricultural University (BAU), Sabour (Bhagalpur), India. He received a doctorate in Horticulture from Banaras Hindu University, Varanasi. He has more than seven years of experience in teaching and research. His research focus is improvement of vegetable crops, especially tomato and okra. Dr. Solankey was awarded the Best Teacher Award and Best Researcher Award in 2016 by BAU, and has twelve other prestigious awards. He has published fifty-five research/review papers, one souvenir paper, six books, one abstract book, and thirty-three book chapters. He is young, dynamic, and wishes to flourish in the field of academia and publications.

Dr. Shirin Akhtar is an enthusiastic and young faculty member at Bihar Agricultural University (BAU), Sabour, Bhagalpur, India. Her field of specialization is vegetable breeding. She received a PhD from Bidhan Chandra Krishi Viswavidyalaya, Mohanpur, West Bengal. Her areas of research are biotic and abiotic stress resistance as well as quality improvement in vegetables, particularly solanaceous crops and okra. She is engaged in teaching undergraduate, postgraduate, and PhD courses and mentoring postgraduate and doctoral students towards new research ideas. She has authored two books, ten book chapters, more than forty research articles in journals of national and international repute, and several popular articles and folders.

Dr. Alejandro Isabel Luna Maldonado received a PhD in Agricultural Sciences from Kyushu University, Fukuoka, Japan, in 2009. He began his career as a lecturer in the Department of Agricultural Engineering at the Autonomous University of Nuevo León in 1992 and was trained in the design and automation of agro-industrial machinery at the Japan International Cooperation Agency. Professor Luna Maldonado became an assistant professor in 1996 and a professor in 2018. He has published thirty-three articles, five book chapters, and four books. He has advised six doctoral theses, five master's theses, and three undergraduate theses. He has served as the head of the educational program of Food Industry Engineering, which has been internationally accredited by Accreditation Board for Engineering and Technology (ABET), since 2009. He has been a member of the Mexican Council of Science and Technology since 2012, and the Program for the Development Teaching Professional (PRODEP) since 2003. He has also been a member of the American Society of Agricultural Engineering since 2012, and the Japanese Society of Agricultural Machinery since 2007.

Dr. Humberto Rodriguez-Fuentes is Professor of Environment and Sustainability at the Autonomous University of Nuevo León, Mexico. He graduated with a doctorate in Agricultural Sciences with a specialty in Water-Soil from the same university. He has forty years of experience in teaching and research. His research is mainly focused in the area of plant factories for the production of highly nutritious vegetables. Since 1990 he has been a national researcher distinguished by the government of Mexico in Biotechnology and Agricultural Sciences. He has published six textbooks, twelve book chapters, and more than fifty articles in journals with strict national/international arbitration. He is also the editor of three books with international distribution.

Juan Antonio Vidales Contreras, MSc, PhD is an agricultural engineer. Since 1985, he has been a full-time professor at the School of Agronomy at the Autonomous University of Nuevo Leon (UANL), Mexico. He received an Agronomy Engineer degree at the same university on 1984. His PhD was awarded by the University of Arizona in 2001. Dr. Vidales Contreras has published more than fifty original research papers in indexed journals, five book chapters, and has participated and contributed in more than twenty scientific meetings.

Julia Mariana Márquez Reyes obtained a PhD in Biotechnology from Autonomous University of Nuevo Leon (UANL), Mexico, in 2013. She specializes in bioreactors with anaerobic activity for the removal of contaminants, phytoremediation for the control of heavy metals in water and soil, enzymatic activity and antioxidant capacity of plant organisms used in environmental biotechnology, and development of sustainable technologies. She began her career as a lecturer of Balance of Matter and Energy, Unit Operations, Environmental Microbiology at UANL. Dr. Márquez Reyes became an assistant professor in 2018. She has been a member of the Mexican Council for Science and Technology since 2009. She has published six scientific papers and one book chapter.

Contents

Preface

The worldwide urban population will double in 30 years, leading to challenges in food and nutritional security as well as environmental problems. The urban population will increase more in developing countries as a result of immigration from rural areas, since people flock to the cities with the expectation of better quality of life there. Looking to accelerated growth in population of cities and small towns, it is expected that by 2050 more than 60 percent of the world's population will live in urban areas. Besides the growing demand for food, there will also be a rapid increase of poverty, unemployment, hunger, and malnutrition in the urban and peri-urban environment, since only 18 percent of the Earth's surface is cultivable or capable of growing plants, while the rest is occupied by seas, mountains, and ice. The little area capable of growing plants (agriculture, horticulture, and green conservation) is highly competed for by housing, industrial and road constructions, and incessant environmental disasters like bush burning, flooding, deforestation, and settlement expansion due to urbanization. The practice of urban horticultural gardening in third world cities to boost food and ornamental plants production, provide job opportunities, mitigate environmental pollution, and promote green space development may bridge these gaps. This is because urban horticulture utilizes the available pieces of land in cities to raise gardens that can be economically productive while contributing to environmental greening.

The chapters in this book cover different components of urban horticulture like utilization of soil and waste materials, implications of automation and robotics systems, nutrient management including fertigation and hydroponics, and the potential and scope of urban horticulture in various parts of the world.

We hope that this publication will be useful for students, professionals, and researchers interested in urban horticulture. We extend our sincere appreciation to the authors, who are from different countries, for their contributions to the book. We thank Ms. Iva Simcic, Commissioning Editor at IntechOpen, for inviting us to be editors of this book. We would like to extend our special appreciation to Author Service Manger Ms. Nina Kalinic Babic for her encouragement and superb support throughout this whole process.

Shashank Shekhar Solankey and Shirin Akhtar
Bihar Agricultural University,
India

Alejandro Isabel Luna Maldonado, Humberto Rodriguez-Fuentes,
Dr. Julia Mariana Márquez Reyes and Juan Antonio Vidales Contreras
Universidad Autónoma de Nuevo León,
Mexico

Section 1

A. Components of Urban Gardens

Chapter 1

Nutrients for Hydroponic Systems in Fruit Crops

Pramod Kumar and Simran Saini

Abstract

Hydroponic systems for crop production are nowadays essential to maximize yields. Sometimes, the benefits of hydroponics have been questioned by the researchers as compared to growing of crops in other soilless culture. The growers raised the crops through hydroponics system get yields more compared to conventional practices as hydroponically grown plants dip their roots directly into nutrient-rich solutions. Therefore, the aim of the current chapter is to provide accurate and updated information about their different nutrients and their composition used hydroponically compared to conventional production mode. This chapter will be divided as the following sections: (1) rationale, (2) nutrient solution technique, and (3) work done on fruit crops. With this chapter, we hope to present an updated information, comparing hydroponic versus conventional technique.

Keywords: soilless culture, hydroponics, conventional production, nutrients, recycling

1. Rationale

Hydroponics is the emerging sector of horticulture that deals with growing of plants in a soilless nutrient solution. This term refers to the use of nutrient and water solution for growing plants without soil. Since the ancient time, this technique is being used from thousands of years that traced from the hanging gardens of Babylon and the floating gardens of China. With the decline in arable land, there is a need of alternative to meet the demand of increasing population, and in this regard, hydroponics serves as an additional channel for crop production. In this technique, the crop plant growth is influenced by certain substances in the water. The German botanists, Julius Von Sachs and Wilhelm Knop developed the first standard formula for the nutrient solutions in 1860–61 where the nutrient solutions contained macronutrients the especially nitrogen, phosphorus, potassium, sulfur, magnesium, and calcium varied concentration depending upon crop. Since 100 years back, William Frederick Gericke popularized the idea that plants could be grown in a solution of nutrients and water. He contributed toward hydroponic culture by producing an effective nutrient solution. In the early 1930s, he did an experiment on production of agricultural crops through nutrient culture and termed it as aquaculture. The term so used was dropped due to culturing of aquatic organisms as aquaculture. During 1930s refinement work on hydroponics was expanded toward Europe, Japan, and North America worked England, Africa, Britain, France, Italy, Spain, and Sweden. In 1937, W.A. Setchell introduced the term "hydroponics." The hydroponic nutrient solution includes minerals in the raw water and nutrients added with fertilizers. The right fertilizer, right dose, and right concentration in the hydroponic nutrient solution greatly depend on the quality of the raw water to be used. This technique has advantages over other methods such as high water use efficiency, improved growth rate, and disease control and also offers more controlled environmental conditions for plants growth and development.

2. Nutrient solution technique

Hydroponics is a technology in which nutrient uptake occurs through plant roots dipped in the nutrient culture. Prior to use of hydroponic culturing, the crop physiological functioning system must be clear. For optimum crop growth and functioning, mixture of sunlight, carbon dioxide, water, and nutrient elements for photosynthetic efficiency is needed. Besides, the minerals are either found naturally in the soil or supplied through the fertilizers to the soil. Thus, it is evident that plant needs mineral derived from the soil for its growth, not the soil which becomes a basic idea behind the development of hydroponics. Moreover, the roots also need an optimum supply of oxygen for uptake and transport of metabolites to the whole plant. In hydroponic system, the roots of plants are in direct contact with the nutrient solution only *vis-a-vis* the absorption of nutrients occurs more easily than soil grown plants. This system also performed the crop plants with faster growth and higher yields and in turn saves energy for extensive root system development and thus, more of energy can be diverted toward leaf and stem growth. Besides, it also offers significant advantages over traditional farming, producing greater yields, faster growth, and possibly year-round crop production. This system further allows recycling of nutrient solution without wastage of water. The amount and composition of the nutrients to which plants have access can be monitored precisely by the grower. It also allows the grower to control the pH level of the solution and protects plants from pests and diseases. However, the literature is still scanty with respect to production of hydroponic fruits [1, 2]. Seedlings of fruit crops can also be grown hydroponically in earlier stages of growth and then transplanted to the field. Different fruit crops namely, grapes, raspberry, blackberry, blueberry, and strawberry can be grown hydroponically. Strawberries and blueberries perform best under hydroponics system because of acidic soil requirement are best suited to blueberries and be grown under hydroponic system as pH level and nutrient content are easily controlled and maintained in hydroponics.

2.1 Hydroponics technique

There are different types of hydroponics technique being employed for growing of plants. There are mainly three types of hydroponic systems.

2.1.1 Nutrient film technique

It is a system in which the nutrient solution is passed through the roots of plants placed in a channel. The plants are placed on channels made up of wood, rigid to flexible tubes or plastic and the nutrient rich solution is either pumped through it or passed under gravitation reaching the root system effectively. This technique is effective only for the plants with large root system. The nutrient film technique (NFT) was developed during the late 1960s by Dr. Allan Cooper at the Glasshouse Crops Research Institute in the U.K. NFT is the growing of plants, bare-rooted in long, waterproof channels, down which flows a very shallow stream of re-circulating water, into which are dissolved all the minerals required to grow healthy plants.

NFT is a hydroponic technique wherein a very shallow stream of water containing all the dissolved nutrients required for plant growth is re-circulated past the roots of plants in a watertight gully, also known as channels. According to the pre-requisite to achieve a nutrient film situation more effectively is described as (i) ensuring the gradient down where water flows is uniform and not subject to localized depressions, (ii) rapid inlet flow rate that a considerable depth of water flows down to the gradient, (iii) adequate width of the channels to avoid any damming up of the nutrients, and (iv) flat channel base but not curved due to which otherwise will be a considerable depth of liquid along the center of a channel with a curved base. N.F.T. system is fairly a simple design. However, this is not best suited for smaller quick growing plants.

2.1.2 Deep film technique

It is a technique in which the plants are grown with roots submerged in floating nutrient solutions (10–20 cm deep) on a flat table. This method is effective for growing plants with short root system and is relatively cheaper.

2.1.3 Substrate method

In this method, the plant growth is supported by using materials such as stones, vermiculite, perlite, etc. in structures such as tubes and pots. This system results in affective utilization of nutrients and reusing of the nutrient solution as the drained solution is re-circulated in the system. This system is commonly utilized in Asia, Europe, and Israel for strawberry cultivation by using several trough systems.

Among the three systems, NFT and DFT are the most commonly used methods. The basic idea behind the working of NFT is the recycling of the nutrient solution. This technique offers major advantages over other systems like low cost of installment, easy operation and management as well as conservation of nutrients and water. In this system, nutrient solution enriched with material like sand, vermiculite or rock wool is passed and re-circulated through a slope consisting of plants placed in a plastic trough. It provides the optimum amount of nutrients to the plant through its root system. It is best suited for short period crops like lettuce, strawberry, and raspberries. Nutrient film technique has been termed as a promising tool in the areas with limited land resources.

Various materials that can be used for hydroponics in fruit crops include that of mineral origin and organic origin. Vermiculite is one of the mineral substances utilized in pears, peach, and tangerine seedlings as it is free of pathogens attack and have high water retention capacity. In case of grapevine, sand because of its easy acquisition is used as it results in the increased absorption of macronutrients. Rice hulls have been found to be effective organic source in growing strawberry plants under hydroponics. For strawberry, perlite or vermiculite is the best growing medium. Materials like coconut coir or peat moss should be avoided as they absorb too much of nutrient solution and cause condition of suffocation to the plants.

For growing strawberries through hydroponics, two systems viz., "closed" and "open" are employed. Among these two, recirculation of nutrient solution occurs in the closed system with plants grown in channels or pipes. The closed system can have continuous nutrient supply or the supply can be at irregular intervals when the plants are grown in pots. In case of pot grown plants, different substrates having high water retention power help in supplying nutrients to the plant system. Whereas in case of open system, there is no recirculation of nutrient solution and is applied to the plants with the help of drippers.

In strawberries, NFT is the most commercially practiced method. In this method, the runners of strawberry are placed in net pots in which the roots are covered with clay medium in the root zone, which help in increasing the strength of plant. They can also be placed with plugs into the net pots which are framed in the NFT channels. It must be placed such that there is continuous contact with the flow of nutrient solution in the initial weeks of root development. The optimum amount of oxygen must also be maintained during the operation of system with 14–16 hours of daylight. The nutrient discharge should be 1–2 liters/minute with circulation pump running all the time. In NFT, it must be grown in low humidity conditions and care should be taken as it is susceptible to root rot.

3. Nutrient solution

Nutrients are the basic elements for hydroponics, and nutrient solution is the liquid fertilizer solution prepared in definite composition to support plant growth. The plants need is fulfilled through the ionic form of nutrients with proper oxygen supply and temperature. Environmental factors and nutrient solution are the two important factors to be considered for productivity in hydroponics. Supply of nutrient elements depends upon the requirement of crop, and the frequency of application is based upon the type and age of crop, the type of material used in media and the prevailing environmental conditions.

The kind of nutrient solution varies according to crop species, their growth stage, environment, and other related factors as there is no ideal nutrient solution available to meet the needs of all the crops. Among fruit crops, a lot of research have been done regarding the nutrient solutions. In grapevine, macronutrient absorption based on nutrient culture of was studied. It was reported that higher accumulation of nutrients resulted in increased vigor of rootstocks Jales, Tropical and Campinas. Solution was also used in pineapple cv. Perola produced through micropropagation in hydroponics system. Long Ashton nutrient solution was used in grapevine under hydroponics [3]. The nutrient solutions for some fruit crops such as peach and pear have not been disclosed.

3.1 Composition

The composition of various nutrients in the nutrient solution plays a major role, as the uptake of these nutrients in optimum amounts affects the functioning of plants, thereby affecting its growth. Testing of water must be done before using it is in hydroponics for nutrient solution to get the accurate details about the properties of water. In strawberry for a closed type of NFT, the nutrient solution with the following composition can be used:

Nutrient elements	Quantity (ppm)
Nitrogen (nitrate form)	160.0
Nitrogen (ammonium form)	15.0
Phosphorus (PO_4)	50.0
Potassium	210.0
Calcium	190.0
Magnesium	50.0
Iron	6.0

Nutrient elements	Quantity (ppm)
Boron	0.50
Manganese	0.50
Copper	0.10
Zinc	0.08
Molybdenum	0.05

There are different nutrient solutions being standardized containing different concentrations of nutrient elements. Hoagland and Arnon nutrient solution has been used for the production of seedlings of guava and pineapple. Furlani et al. nutrient solution was used in production of guava seedlings. Yamazaki solution can be used for strawberry which includes N(NO_3:5; NH_4:0.5); P:1.5; K:3; Ca:2; Mg:1; S:1; Fe:3; B:0.5; Mn:0.5; Zn:0.05; Cu:0.02; Mo:0.01 in meq/L [4].

Nutrient elements	Hoagland and Arnon (1938)	Furlani et al. (1999)	Hewitt [3]	Cooper [5]
		mg/L		
N	210	202	168	200–236
P	31	31	41	60
K	234	193	156	300
Ca	160	142	160	170–185
Mg	34	39	36	50
S	64	52	48	68
Fe	2.5	0.26	2.8	12
Cu	0.02	0.04	0.064	0.1
Zn	0.05	1.8	0.065	0.1
Mn	0.5	0.37	0.54	2.0
B	0.5	0.06	0.54	0.3
Mo	0.01	0.11	0.04	0.2

3.2 Nutrient preparation

The nutrient solution for hydroponics can be either bought premixed or can be prepared by self. Plants require same macro and micronutrients but in different ratios. So, the nutrients supplying fertilizers must be bought based on the plants' need as each nutrient has distinct function in different plants. For the preparation of nutrient solution, the nutrient fertilizers are mixed with water which breaks downs to release nutrients. The selection of fertilizers should be such that the amount of water and nutrients present in the solution are equal to the amount of water and nutrients taken up by the plants. The nutrient solution preparation requires water of good quality which is contamination free for which chemical analysis is important. The formulations must be based upon the targeted crop and must supply all the essential nutrients. The nutrient solution preparation also requires maintaining optimum levels of pH and EC. For the proper growth of strawberry plants, the pH of nutrient solution used must be between 5.5 and 6 and the ideal EC range is 1.8–2.0 dS/m during growth period and 1.8–2.5 dS/m during fruiting stage. The pH of the solution can be maintained through potassium hydroxide used to increase pH or through phosphoric acid to lower down the pH. At EC higher than 1.2 dS/m to prevent the damage to strawberry plant root area of plants can be flushed with clean water for the removal of accumulated salts.

4. Recycling of nutrient solution

To meet the requirement of plant without any loss of nutrients, the closed system of hydroponics offers a huge benefit for recycling of nutrients reducing economic as well as environmental costs. In a closed system, the water and nutrient supply is equal to their quantity taken up by the plants [6]. This system provides controlled nutrient supply with minimal leaching losses and reduced environmental contamination. There is a continuous supply of nutrient solution touching the roots of the plant which after passing down is recirculated and is again available for the use of the plants. Among the added fertilizers, only 50% is utilized by the plant and 70% of the added water is utilized by the plant for its proper growth and transpiration [7]. Recycling of nutrients through closed system of hydroponics and nutrient film techniques is very efficient as it uses only 10% of the water and 25% of fertilizer to that of conventional systems. The mineral content of the added fertilizers may be reduced due to the uptake by plants which may be replenished from time to time. The recycling and reusing of nutrients and water having huge advantages also poses some issues. There is increase in EC of the nutrient solution if water uptake is greater than the nutrients and reduced EC due to greater nutrient uptakes which as a result disrupt the recycling mechanism of the system. There may be problem of increased concentration of salts, toxic ions, and pathogens in the nutrient solution where recycling of pathogens occurs along with the solution in the system resulting in their build up. The problems being faced in the closed system can be removed by using ultraviolet treatment, heat treatment, and slow sand filtration. It has been proved that among the previously mentioned, slow sand filtration is best as it is chemical free, easy to maintain, and energy efficient with adaptability in components [8]. Bio sand filter is used in the system against pathogens of *Pythium* spp. and *Phytophthora* spp. In case of open system, the nutrient solution is not recirculated but released into the environment after the crop cultivation. It has been proposed that the nutrient solution released to environment can be recycled to be used as irrigation water without by reducing further pollution chances [9].

5. Frequency of application

For growing strawberry hydroponically, the plants must be fed with nutrient solution daily and best time being 6:00 am–8:00 am. The application must be such that conditions like overwatering and drying not occur. For the early stages of plant growth when the plant is small, less amount of nutrients are required. In actively growing period and during summers, large amount of water is consumed by the plants due to increased transpiration rate. The water requirement also varies depending upon the environmental conditions maintained. The amount of nutrients to be added to the system must be based upon the crop usage. The quantity of water and nutrients taken up by the plants can be known by measuring the EC daily. Lower amount of EC indicates more nutrient uptake, and higher level indicates increased water uptake. Based upon the EC levels, the water must be added to avoid buildup of salts. The reservoir must be filled with water once it is lower than the required volume and must be checked for EC and pH. If the EC and pH of the solution is not stable as per the requirement, it must be adjusted accordingly. The refilling of reservoir from time to may result in imbalance of nutrients in the nutrient solution. So, it must be dumped after a period of time to avoid any interruption in the growth of pant. If the reservoir is small, the solution must be dumped after every 10–15 days while in case of large reservoirs once a month.

6. Nutrient need through hydroponics

In soil grown plants, the fertilizers are added into the soil or applied through foliar application, but in hydroponics, a solution of ionic compounds helps in delivering nutrients into the plant system. Hydroponics is better in meeting plants nutrient need as under the manipulated set of conditions nutrients are directly supplied to the roots by coming in contact with them. Hydroponics, due to their better control over the environmental conditions, has been proved to be superior and sustainable for growing of different crops. Particularly in case of berry crops like strawberry, it has turned to be a very effective method producing fruits of superior quality with high yielding potential. Hydroponically, grown strawberries have been found to produce fruits with higher amount of vitamin C, vitamin E, and total polyphenols. By following different systems in hydroponics and different substrates, the nutrient demand of the plant can be met more efficiently. The supply of optimum amount of nutrients and water must be ensured depending upon the crop need so that the plant continues to grow without any lack or excess of both nutrients and water. With several discussed advantages, hydroponics is better choice over conventional methods to produce fruits with reduced water and fertilizer use.

7. Work done

Maximized growth and yield with mixture of perlite (60–80%) and peat (20–40%) in strawberry [10]. Maximum yield was recorded in strawberry grown in perlite mixed with coco coir or vermiculite in vertical hydroponic system [11]. Takeda [12] suggested that transplant plug plants were superior in increasing yield to fresh plants for hydroponic production of strawberry cv. Sweet Charlie and Camarosa. Costa et al. [13] concluded that the carbonized rice husk substrate produced more than one crop (off-season) in soilless culture in strawberry cv. Albion frigo. Treftz et al. [14] reported combined benefits of environment and better sensory attributes, and it is desirable to grow strawberry hydroponically. Treftz and Omaye [15] noted that growing strawberries in hydroponic systems are more sustainable and superior to soil grown systems. Ramirez-Gomez et al. [16] reported maximum yield with vertical hydroponic pots system; the maximum number of fruits with vertical four pipes system and inferior quality fruit were produced with vertical three pipes system in strawberry.

Ramirez-Arias et al. [17] reported maximum yield with vertical hydroponics system and the lowest was found in three level horizontal systems in strawberry cv. Festival. Peralbo et al. [18] concluded that maximum yield was produced by peat as compared to cork compost in both open and closed hydroponics system in strawberry. Miranda et al. [19] found that closed hydroponic system of gutters and grow bags was superior than the open system in saving water and fertilizer in strawberry. Roosta and Afsharipoor [20] concluded that dry weight, leaf area, number of runners, Leaf N, P, K, Fe, Mg, and yield was significantly higher in hydroponics as compared to aquaponics except for soil perlite. Portela et al. [21] noted higher yield through nutrient solution between EC ranges of 1.2–1.5 dS/m in NFT hydroponics system in strawberry cv. Camarosa. Vikas et al. [22] reported maximum plant height and maximum number of fruits with sewage sludge and cocopeat (20:80), whereas the maximum number of leaves and yield was observed with sewage sludge and cocopeat (30:70) in strawberry under hydroponics. Choi et al. [23] concluded that FAI technique for coir substrate was best in hydroponics due to sustainable use of water and fertilizers in strawberry. Albaho et al. [24] concluded that continuous sub irrigation capillary system is the best among hydroponics in strawberry. Jun et al. [4] reported that nutrient solution with EC ranges between 0.8 and 1.2 dS/m during low temperature season in hydroponically grown strawberry cv. Maehyang.

Lee et al. [25] noted that nutrient solution with EC of 1.0 dS/m is best for hydroponically produced strawberry cv. Albion and Goha. Andriolo et al. [26] reported maximum fruit yield with EC 0.9 dS/m under closed soilless growing system in strawberry. El-Sayed et al. [27] noted significant improvement in vegetative growth characters, leaf chemical content, and yield in perlite: peat moss substrate under hydroponics in strawberry cv. Festival. Ebrahimi et al. [28] reported maximum number of fruits and yield with cocopeat + perlite substrate and improved quality with peat + sand + perlite substrate in strawberry cv. Camarosa and Selva. Marinou et al. [29] concluded that sawdust was best substrate medium under hydroponics in strawberry. Caruso et al. [30] reported improved fruit quality through nutrient solution with EC 1.3 dS/cm in spring season and through 2.2 mS/cm in winters under NFT in strawberry cv. Alpine. Souza et al. [31] observed fastest transplanting stage and grafting stage at 30 and 61 days after transplanting under hydroponics system for commercial grafts production in peach. Motosugi et al. [32] reported increase in anthocyanin level with ammonium nitrogen nutrient solution at pH 3–3.5 under NFT in grapevines.

Author details

Pramod Kumar* and Simran Saini
YS Parmar University of Horticulture and Forestry (Fruit Science), Solan, Himachal Pradesh, India

*Address all correspondence to: pk09sharma@rediffmail.com

References

[1] Buchanan DN, Omaye ST. Comparative study of ascorbic acid and tocopherol concentrations in hydroponic- and soil-grown lettuces. Food and Nutrition Sciences. 2013;**4**:1047-1053

[2] Selma MV, Luna MC, Martinez-Sanchez A, Tudela JA, Beltran D, Baixauli C, et al. Sensory quality, bioactive constituents and micro-biological quality of green and red fresh-cut lettuces (*Lactuca sativa* L.) are influenced by soil and soilless agricultural production systems. Postharvest Biology and Technology. 2012;**63**:16-24

[3] Hewitt EJ. Sand and Water Culture Methods Used in the Study of Plant Nutrition. 2nd ed. Farnham Royal, England: Commonwealth Agricultural Bureaux; 1966. 547p

[4] Jun HJ, Byun MS, Liu SS, Jeon EH, Lee YB. Effect of nutrient solution strength on growth, fruit quality and yield of strawberry 'Maehyang' in hydroponics. Korean Journal of Horticultural Science and Technology. 2013;**31**:173-178

[5] Cooper A. The ABC of NFT: Nutrient Film Technique: The World's First Method of Crop Production Without a Solid Rooting Medium. London: Grower Books; 1979. 181p

[6] Voogt W, Sonneveld C. Nutrient management in closed growing systems for greenhouse production. In: Goto et al, editors. Plant Production in Closed Ecosystems. Kluwer Academic Press; 1996. pp. 83-102

[7] Choi SY, Lee YB, Kim JY. Nutrient uptake, growth and yield of cucumber cultivated with different growing substrates under a closed and an open system. Acta Horticulture. 2001;**548**:543-550

[8] Barth G. Slow Sand Filtration. Extracts of Horticulture Research and Development Corporation. 1998. Available from: http://www.sardi.sa.gov.au/pages/horticulture/ofc/hort_ofc_slowsandfiltration.htm:sectID=119&tempID=71

[9] Jensen MH, Collins WL. Hydroponic vegetable production. Horticultural Reviews. 1985;**7**:483-558

[10] Anagnostou K, Vasilakakis MD. Effect of substrate and cultivar on earliness, plant productivity, and fruit quality of strawberry. Acta Horticulturae. 1995;**379**:267-274

[11] Wortman SE, Douglass MS, Kindhart JD. Cultivar, growing media, and nutrient source influence strawberry yield in a vertical, hydroponic, high tunnel system. HortTechnology. 2016;**26**:466-473

[12] Takeda F. Strawberry production in soilless culture systems. Acta Horticulturae. 1999;**481**:289-295

[13] Costa RC, Calvete EO, Mendonça HFC, Campagnolo A, Chiomento JLT. Performance of day-neutral strawberry cultivars in soilless culture. Australian Journal of Crop Science. 2016;**10**:94-100

[14] Treftz C, Zhang F, Omaye ST. Comparison between hydroponics and soil-grown strawberries: Sensory attributes and correlations with nutrient content. Food and Nutrition Sciences. 2015;**6**:1371-1380

[15] Treftz C, Omaye ST. Nutrient analysis of soil and soilless strawberries and raspberries grown in a greenhouse. Food and Nutrition Sciences. 2015;**6**:805-815

[16] Ramirez-Gomez H, Sandoval-Villa M, Carrillo-Salazar A, Muratalla-Lua A.

Comparison of hydroponic systems in the strawberry production. Acta Horticulturae. 2012;**947**:165-172

[17] Ramirez-Arias JA, Hernandez - Ibarra U, Pineda- Pineda J, Fitz-Rodriguez E. Horizontal and vertical hydroponic systems for strawberry production at high densities. Acta Horticulturae. 2018;**1227**:331-338

[18] Peralbo A, Flores F, Lopez-Medina J. Recirculating nutrient solution in strawberry. Acta Horticulturae. 2005;**697**:101-106

[19] Miranda FR, Silva VB, Santos FSR, Rossetti AG, Silva CFB. Production of strawberry cultivars in closed hydroponic systems and coconut fibre substrate. Revista Ciencia Agronomica. 2014;**45**:833-841

[20] Roosta HR, Afsharipoor S. Effects of different cultivation media on vegetative growth, ecophysiological traits and nutrients concentration in strawberry under hydroponic and aquaponic cultivation systems. Advances in Environmental Biology. 2012;**6**:543-555

[21] Portela IP, Peil RMN, Rombaldi CS. Effect of nutrient concentration on growth, yield and quality of straw-berries in hydroponic system. Horticultura Brasileira. 2012;**30**:266-273

[22] Vikas KA, Kumar A, Singh A, Kumar P, Rafie J, Verma P, et al. Influence of different nutrient concentration on strawberry under hydroponic cultivation system. International Journal of Current Microbiology and Applied Sciences. 2017;**6**:2999-3006

[23] Choi KY, Choi EY, Kim IS, Lee YB. Improving water and fertilizer use efficiency during the production of strawberry in coir substrate hydroponics using a FDR sensor automated irrigation system. Horticulture, Environment and Biotechnology. 2016;**57**:431-439

[24] Albaho M, Thomas B, Christopher A. Evaluation of hydroponic techniques on growth and productivity of greenhouse grown bell pepper and strawberry. International Journal of Vegetable Science. 2008;**14**:23-40

[25] Lee YH, Yoon CS, Park NI, Yeoung YR. Influence of various nutrient concentrations on the growth and yield of summer strawberry cultivars cultivated in a hydroponic system. Horticulture, Environment and Biotechnology. 2015;**56**:421-426

[26] Andriolo JL, Janisch DI, Schmitt OJ, Vaz MAB, Cardoso FL, Erpen L. Nutrient solution concentration on plant growth, fruit yield and quality of strawberry crop. Ciencia Rural. 2009;**39**:684-690

[27] El-Sayed SF, Hassan HA, Abul-Soud M, Gad DAM. Effect of substrate mixtures and nutrient solution sources on strawberry plants under closed hydroponic system. International Journal of Product Development. 2016;**21**:97-127

[28] Ebrahimi R, Souri MK, Ebrahimi F, Ahmadizadeh M. Growth and yield of strawberries under different potassium concentrations of hydroponic system in three substrates. World Applied Sciences Journal. 2012;**16**:1380-1386

[29] Marinou E, Chrysargyris A, Tzortzakis N. Use of sawdust, coco soil and pumice in hydroponically grown strawberry. Plant, Soil and Environment. 2013;**59**:452-459

[30] Caruso G, Villari G, Melchionna G, Conti S. Effects of cultural cycles and nutrient solutions on plant growth, yield and fruit quality of alpine strawberry (*Fragaria vesca* L.) grown in hydroponics. Scientia Horticulturae. 2011;**129**:479-485

[31] Souza AG, Chalfun NNJ, Faquin V, Souz AA. Production of peach grafts

under hydroponic conditions. Ciencia e Agrotecnologia. 2011;**35**:322-326

[32] Motosugi H, Beppu K, Sugiura A. Studies on NFT culture of 'Kyoho' grapes: Water stress symptoms and fruit quality of grapevines supplied ammonium nitrogen at low pH after veraison. Acta Horticulturae. 1995;**396**:227-234

Chapter 2

Hydroponic Systems for Arabidopsis Extended to Crop Plants

Lawrence Griffing and Krishna Kumar

Abstract

When using Arabidopsis grown hydroponically for gene and drug discovery, a method for translating this approach to crop (and weed) species needs articulation and investigation. In this review, we describe existing inexpensive, frequently aseptic, hydroponic systems for Arabidopsis and compare them to other hydroponic methods for gene and drug discovery in crop plants. Besides gene and drug discovery, an important use of hydroponic analysis is for understanding growth in controlled, enclosed systems, such as during spaceflight and in simulated extraterrestrial environments. When done initially with Arabidopsis, will these results apply to the growth of other species? We highlight the strengths and weaknesses of existing translational hydroponic approaches whereby results with Arabidopsis extend to other plant species. We find that the existing or slightly modified hydroponic approaches used in Arabidopsis research extend well to crop plants that grow upright about 40 cm in height, e.g., monocots, such as rice, and dicots, such as soybean. However, other, taller species such as maize, or vining species such as tomato, require extensive modification to provide larger enclosures and root stabilization.

Keywords: translational research, drug discovery, herbicide discovery, gene discovery, bioregenerative systems, speed breeding, fast generation cycling systems

1. Introduction

Arabidopsis thaliana (hereafter referred to as Arabidopsis) is a model plant system and, unlike most other plants, has a very large number of sequenced chemically induced mutations and libraries of insertional mutations in genes of known and unknown function [1]. This genetic power of Arabidopsis makes it a continuing resource for studying the functions of genes under a variety of conditions. Often, these conditions are best-controlled using hydroponic systems to control nutrients or other abiotic (e.g., drugs, light, solute stress) or biotic (microbes) interactions in the rhizosphere. Furthermore, with tight control of rhizosphere conditions using hydroponics, other experiments on the shoots, leaves or flowers can proceed.

2. Hydroponic methods for Arabidopsis

Tocquin et al. [2] briefly review earlier approaches to Arabidopsis hydroponics. More recent studies have developed low cost, efficient systems, which are based on the use of reusable and sometimes sterilizable plastic materials [3–7]. These systems differ in whether they offer (a) aseptic conditions (b) synchronous, rapid growth, (c) growth to maturity and (d) low cost for set-up and maintenance. All of them grow the plants in simple, defined liquid media (usually a ½-strength Murashige and Skoog medium or nitrogen-supplemented ¼-strength Hoaglands solution). They also provide access to the rhizosphere for root phenotyping and drug delivery. As shown in **Figures 1**, **2**, the root system is readily available for imaging or for biochemical analysis. Efficient harvest of intact roots is difficult in soil-grown plants, whilst hydroponics provides clean, and potentially aseptic, harvest of roots. However, hydroponics depresses the formation of root hairs and produces developmental changes in other root tissues (reviewed in [8]).

Tocquin et al. [2] and Monte-Bello et al. [7] describe systems where the plants grow in agar-filled end-clipped mini-tubes placed either in the holes of an autoclavable pipette tip holder [7], or in holes drilled into dark plastic sheets covering an opaque plastic bin [2], **Figure 1**. The setup by Tocquin et al. [2] is not aseptic and uses polyethylene plastic, which is not autoclavable, but does provide synchronous, rapid growth using optimized nutrients based on the modifications of Hoagland's medium [9]. Although not aseptic, the dark plastic sheet discourages the growth of mold and algae at the surface. The media is not circulating or artificially aerated, known in the popular literature as Kratky-type hydroponics [10]. Groups of a dozen or so plants grow in a single tray and the equipment is scalable to larger plantings (trays) or to larger plants (see below). By lifting the plastic sheet, the roots are easily harvested, **Figure 1**. The cost is low, but the cut and drilled plastic sheets are not available commercially. Unlike the similar system [7], it provides media and space required to grow the plants to maturity. Arabidopsis has a shorter generation time in hydroponics than in soil [2], where single-pulse long day lighting induces flowering in 6–7 weeks with hydroponics and 8 weeks in soil [11]. On the other hand, the Monte-Bello setup provides only enough room to grow plants hydroponically to a 4-leaf stage (3–4 weeks), but under aseptic

Figure 1.
Tray-based hydroponics for Arabidopsis. (A) Top view of plants grown by technique of Tocqui et al. [2]. (B) Side-view of root systems grown [2]. (C) Top view of plants grown in pipette holders by the technique of Monte-Bello et al. [7]. (D) Side view of plants grown by Monte-Bello et al. [7]. A&B from Tocquin et al. [2], scale bar = 3 cm. C&D from Monte-Bello et al. [7].

Figure 2.
Cup-based hydroponics for Arabidopsis. (A) Seeds sown on mesh, inset = magnified view of Arabidopsis seed on mesh. (B) Top view of 12-day-old seedlings. (C) Side view of 12-day-old seedlings. (D) Top view of 21-day-old seedlings. (E) Side view of 21-day-old seedlings [5].

conditions, **Figure 1**. It is unclear whether, at this stage, it is feasible to transfer plants to other, larger tube systems [4].

Other Arabidopsis hydroponic systems use an insert into a plastic box or cup [3, 5, 6, 12]. All produce synchronous growth to maturity. The plastic cup system [5] is autoclavable when using cups of polypropylene. Covering the plants with an autoclavable lid or, in later stages, a tall cup, maintains sterile culture in early stages of growth. The polypropylene cups are very cheap, because they are available commercially as single-use plastic containers. In this cup system, plants germinate on a screen (plus agar with medium) wedged between a smaller upper cup and larger lower cup, **Figure 2**. Lifting out the screen makes the root system available for analysis and harvest, as is also described in the non-sterile hydroponic culture of Arabidopsis on a supported nylon screen in a beaker [13].

Arteca and Arteca [12] and Nguyen et al. [6] use classic Magenta™ GA-7 boxes as the media chamber and float foam squares containing the plants on the surface of the media. These are modifications of one of the earliest hydroponic culture systems, consisting of a water or nutrient reservoir, an air pump, tube, and a floating platform [14, 15]. Nguyen et al. [6] aseptically pre-germinated the plants on agar and then gently wedged them into foam holders. Robison et al. [3] report another version of this using rock wool and inexpensive food container boxes. More handling of the delicate plants occurs when there is transplantation of plants initially grown in agar. Germination directly on rock wool plugs (with a 0.15% w/v agar) is also possible [16]. Once in the foam or rock wool holders and open to the environment, plant growth is no longer aseptic. One procedure [6] also includes aeration of the media with a bubble stone. Others [12] show that there is no effect of bubble stone aeration on growth.

3. Closed and semi-closed systems for Arabidopsis

Some hydroponic systems are open systems that add new media and do not reuse or recycle old media [17]. Providing a continuous supply of defined nutrient or drug-containing solution makes open systems costly and does not take advantage of the ease with which hydroponic nutrients can be recycled or continuously re-used. In several of the above Kratky-type methods for Arabidopsis hydroponics, there is regular, but infrequent (weekly) modification of the nutrient solution and they provide a semi-closed, constant, uncycled medium. Closed and semi-closed systems that use flow, such as the nutrient film technique (NFT), deep flow technique (DFT) and aeroponics (misted nutrient solution sprayed on the roots), cycle a larger volume of nutrient medium. However, both cycled and uncycled hydroponics share the problem that the ionic balance or salinity within the nutrient medium can change over time [18, 19]. Computational approaches based on continuous read-out from ion-selective electrodes provide real-time optimization of hydroponic media [20–22]. Such optimization in closed systems require, as part the enclosure, additives or scrubbers (such as ion exchange resins) to add or remove certain ions or nutrients.

Semi-closed and closed, aseptic systems for Arabidopsis growth are useful for (1) drug discovery, (2) gene discovery, (3) plant-microbe interactions, and (4) growth in non-terrestrial environments. The advantage of closed or semi-closed hydroponic systems for discovery of drugs effective against plants is that, unlike the situation in animals where target animals are not available for ethical or logistical reasons, it provides a method to evaluate the living target organism from early to late stages of growth [23]. If done on sufficient scale, it can be a form of high throughput in vivo screening. Whole organism screening covers important drug or herbicide properties such as uptake, efficacy, and breakdown. However, the frequent use of Arabidopsis, and other model organisms, e.g., duckweed (*Lemna* spp.) and the model grass, *Brachypodium distachyon*, as test species for drug discovery, particularly for herbicides, is problematic [24]. Model species may not have the same mechanisms for uptake, delivery, and metabolism of the drugs as agricultural weeds and crops. They have different life cycles and environmental preferences and constraints. Therefore, it is important to extend the work on closed and semi-closed systems using Arabidopsis to crop species and weed species, as described below.

This is also true for gene discovery. Identifying and preliminary mapping of multi-genic quantitative trait loci (QTLs) that produce desirable phenotypes is possible with recombinant inbred lines (RILs) and near isogenic lines (NILs) of Arabidopsis, if there is minimal environmental contribution to the character. The standard conditions of closed or semi-closed hydroponics seem ideally suited for these studies. When soil is used, it could be a source of irreproducibility. Growth conditions using soils produce irreproducible Arabidopsis leaf phenotypes in different labs, even controlling for many environmental variables and nutrient conditions [25]. Similar multi-lab reproducibility experiments with hydroponically grown Arabidopsis plants are not available, but provide an exciting opportunity for new study. Exploiting closed or semi-closed hydroponics tested with Arabidopsis could be an important step in speed breeding and the analysis of RILs or NILs of fast-cycling crop species [26], as described below.

Sequencing-based analysis of bacterial communities on plants reveals the diversity and complexity of the interaction of plants with the microflora of the rhizosphere [27]. Hydroponic approaches to analysis of Arabidopsis-microbe interactions provide a way to monitor how multiple bacterial species colonize the root or interact with each other to form these complex interactions [28]. These approaches require aseptic culture of Arabidopsis and controlled introduction of monocultures of bacteria into the media. Harris et al. [28] adapted the simple, inexpensive, closed hydroponic system described above [5] to analyze the colonization of Arabidopsis roots with *Pseudomonas*, *Arthrobacter*, *Curtobacterium*, and *Microbacterium* species.

Completely enclosed, but not necessarily hydroponic, systems were common for early studies on the growth of plants in extraterrestrial environments (see review, [29]). The cultivation system of choice is recirculating, enclosed hydroponics, however, for future space flights that would include plants in a life support system [30]. The new recirculating enclosed system by NASA and ORBITEC, however, has Arcelite as a root support matrix with porous tubes to pump the nutrient solution [31].

Figure 3.
Advanced plant habitat of NASA showing growing wheat plants in an enclosed chamber with an Arcelite substrate (https://www.nasa.gov/sites/de-fault/files/atoms/files/ad-vanced-plant-habitat.pdf).

The initial tests of the Plant Habitat-01 in the International Space Station will be studies on Arabidopsis. Future experiments will include durum wheat, as shown in **Figure 3**. Studies on soil-less bioregenerative life support systems funded by the European Space Agency use the nutrient film technique (NFT) of hydroponics [32], with the caveat that the implementation of such a technique in microgravity is yet to come. All of these life support systems are gas-tight enclosures that will monitor gases emitted by the plants, because the recycling of carbon dioxide and oxygen by plants or other photosynthetic organisms will be a necessity in long-term flights such as those to Mars.

4. Extending closed and semi-closed systems of Arabidopsis to crop species

A semi-closed hydroponic system that is very close to those described above for Arabidopsis is the single-tube hydroponics of Kuroda and Ikenaga [33]. As in the procedures for Arabidopsis used by Nguyen et al. [6] and Robison et al. [3], the plants initially germinate on an agar (gelrite) medium containing ½ strength Murashige and Skoog medium. They grow a variety of crop plants, i.e., rice, soybean, Azuki beans, and corn, instead of Arabidopsis. At 2 weeks, Kuroda and Ikenaga [33] transplant the intact germinated seedlings into 12 ml polypropylene tubes with two holes cut into the sides to allow the entry and exit of hydroponic medium (1/10 strength Murashige and Skoog medium). A covered outer tray contains the medium and a rack for the tubes in which the plants grow. The tube supports the plant during culture and contains the root ball of each plant, thereby facilitating removal for analysis without damaging the roots. The size of the culture tube is larger than that used by Tocquin et al. [2] for Arabidopsis because the seeds and new roots of crop plants are much larger, but the tray system for growth in hydroponics is very similar. For soybean, an additional prop supports the plant during growth. Rice and soybean plants grown in single tube hydroponics produce high viability seed with seed weights equal to or exceeding plants grown in the field. Single tube hydroponics facilitate analyzing and screening the T1 seeds from the transgenic plants with shorter generation times and small amounts of seeds.

The hydroponic system of Conn et al. [4] directly translates an Arabidopsis hydroponic culture system to crop plants. They use a system of hole-punched plastic trays to start the plant, followed by transplantation of the seedlings into a plastic tube that can be set in a larger tray with aeration. The 50 ml tube in this case, although it did not confine the root as much those used by Kuroda and Ikenaga [33], kept the roots of separate plants free from tangling, thereby facilitating measurement and analysis. Although Arabidopsis was the main test plant for this system, wheat, cucumber, and tobacco also successfully grew.

Although both techniques [4, 33] require transplantation of newly germinated plants, which has the downside of more manipulation, transplantation may be desirable for crop plants with low germination rates and for studies on post-emergent drug treatments. It has the additional advantage of protecting the plant from water molds and other contaminants because the initial germination is aseptic.

It becomes apparent in studies that translate work on hydroponically grown Arabidopsis to crop plants that just the difference in physical size of the seeds and plants dictates some of the modifications. In contrast to Arabidopsis, experiments involving larger plants require root stabilization. When larger crops, such as *Zea mays*, grow in hydroponic conditions, lack of support for the root system can result in breakage and damage of the lateral root system [34]. When growing wheat varieties to test the effects of salinity, Munns and James [35] used quartz rock as a stabilizing substrate in a hydroponic flow system. More complex, but definable, substrates may be necessary because they interact with nutrients and help determine their availability, e.g., a defined clay substrate for corn [36, 37].

The generation time for soil-grown Arabidopsis decreases by 1 to 2 weeks when grown hydroponically on defined medium [2, 11]. Accelerated breeding programs for crops facilitate the production of RILs and NILs for gene discovery. In fast generation cycling systems, **Figure 4** [26], plants with long generation times, such as crop plants, are sped up using a variety of technologies. Speed breeding can produce generation times that are a third to a half the time [38, 39]. One of the technologies used in speed breeding is in vitro growth. With a neutral rhizosphere support medium, such as agar for Arabidopsis [40], speed breeding for Arabidopsis translates to speed breeding in wheat [41]. Besides achieving fast growth of the seedling, an important feature of many fast generation cycling systems is overcoming seed dormancy with early stage embryo culture in vitro, **Figure 4** [26, 42] or harvesting immature seed and drying it [38, 39]. With embryo culture, there will always be aseptic transplantation of agar-grown embryos, but for immature seeds, transplantation is not necessary.

Complete enclosure of the growing crop is one of the features of recent speed breeding technologies [38, 39, 43]. Soybeans grown hydroponically using NFT in the completely enclosed bioregenerative life support systems have a 110–133 day

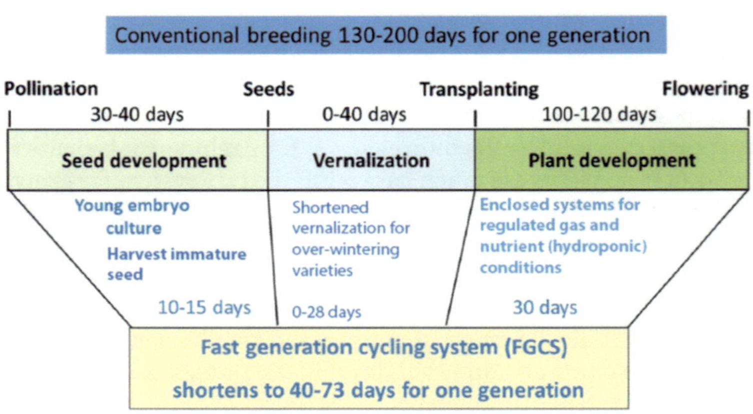

Figure 4.
Schedules for speed breeding in fast generation cycling systems showing the timesaving steps to reduce generation times (adapted from [26]).

generation time [32]. Soil grown soybeans have a generation time of 132 days, but growing them in an enclosed system with elevated carbon dioxide decreases the generation time to 70 days [43]. Combining hydroponics of crop species with other technologies of speed breeding might produce even shorter generation times or higher yields and seed viability.

Complete enclosure of the growing crop is also a feature of crops grown for space exploration. Because soybean is one of the species best suited for growth in space [44], its nitrogen fixing symbiosis with bacteria is of interest. As described above, hydroponic submersion systems inhibit root hair development, upon which the initial stages of nodule formation depend. Hence, it is not a surprise to see that hydroponic inoculation of soybean with its nodulation partner, *Bradyrhizobium japonicum*, does not improve nitrogen use efficiency [44]. However, other kinds of plant growth promoting microbes (PGPMs), including some of those tested in hydroponic systems with Arabidopsis [28], produce higher photosystem II efficiency in hydroponically-grown soybean plants [45]. This could be beneficial in speed breeding, which improves with improved photosynthesis achieved with elevated carbon dioxide [43]. Those plants grown to maturity in enclosed hydroponics and inoculated with PGPMs show stabilized microbial communities over time [46].

5. Conclusions: translational research on hydroponics from Arabidopsis to crops

As described in Woodward and Bartel [1], research on Arabidopsis can sometimes directly translate into discoveries in crops. One example that they use is the expression of MYB12 in tomatoes, which derived from initial discoveries in Arabidopsis revealing increased production of flavonoids upon overexpression. The overexpression of MYB12 in tomatoes produces so much flavonoids, the color of the fruit changes from red to orange [47]. However, this small mustard family plant has a growth habit and life cycle so different from most crop plants, can lessons learned from hydroponic studies on Arabidopsis be translated to crops? The answer is: mostly. Most of the technical approaches used with Arabidopsis translate to crop plants with minor modification, except for those crops that are very large and need extra support for growth. The benefits of using Arabidopsis for investigating the different techniques of hydroponics are those that make it valued as a model organism, i.e., its size, well-characterized genome, and short generation time. In fact, given the depth of knowledge on gene function in Arabidopsis, current research on Arabidopsis hydroponics could apply more widely to studies on fast breeding crop plants for gene discovery, on target plants for herbicide and drug discovery, and on plants used for bioregenerative life support systems in space.

Adoption of some the techniques used in Arabidopsis hydroponics could decrease the cost and size (important for space studies) of enclosed test systems without changing the viability and yield of the crop plants grown in those systems. For example, the effects of space travel are varied and complicated. However, most of the work done to date has focused on the microgravity component of space flight without the proper control of having a 1-g set of plants growing in the same space vessel [29]. Because Arabidopsis is small and well characterized, the initial tests for the design and implementation of these proper controls may be more feasible (and the data achieved more insightful) for Arabidopsis than for the crops identified as "the best" space plants, i.e., durum and bread wheat, soybean, and potato. Once done with Arabidopsis, the work would translate to these other species.

Acknowledgements

The authors acknowledge the support of Biology Department, Texas A&M University.

Author details

Lawrence Griffing* and Krishna Kumar
Biology Department, Texas A&M University, College Station, TX, United States

*Address all correspondence to: griffing@tamu.edu

References

[1] Woodward AW, Bartel B. Biology in bloom: A primer on the Arabidopsis thaliana model system. Genetics. 2018;**208**(4):1337-1349

[2] Tocquin P, Corbesier L, Havelange A, Pieltain A, Kurtem E, Bernier G, et al. A novel high efficiency, low maintenance, hydroponic system for synchronous growth and flowering of Arabidopsis thaliana. BMC Plant Biology. 2003;**3**:2

[3] Robison MM, Smid MPL, Wolyn DJ. High-quality and homogeneous Arabidopsis thaliana plants from a simple and inexpensive method of hydroponic cultivation. Canadian Journal of Botany. 2006;**84**(6):1009-1012

[4] Conn SJ, Hocking B, Dayod M, Xu B, Athman A, Henderson S, et al. Protocol: Optimising hydroponic growth systems for nutritional and physiological analysis of Arabidopsis thaliana and other plants. Plant Methods. 2013;**9**(1):4

[5] Alatorre-Cobos F, Calderon-Vazquez C, Ibarra-Laclette E, Yong-Villalobos L, Perez-Torres CA, Oropeza-Aburto A, et al. An improved, low-cost, hydroponic system for growing Arabidopsis and other plant species under aseptic conditions. BMC Plant Biology. 2014;**14**:69

[6] Nguyen NT, McInturf SA, Mendoza-Cozatl DG. Hydroponics: A versatile system to study nutrient allocation and plant responses to nutrient availability and exposure to toxic elements. Journal of Visualized Experiments. 2016;**113**:e54317. DOI: 10.3791/54317

[7] Monte-Bello CC, Araujo EF, Martins MCM, Mafra V, da Silva VCH, Celente V, et al. A flexible low cost hydroponic system for assessing plant responses to small molecules in sterile conditions. Journal of Visualized Experiments. 2018;**138**:e57800. DOI: 10.3791/57800

[8] Shavrukov Y, Genc Y, Hayes J. The use of hydroponics in abiotic stress tolerance. In: Asao DT, editor. Hydroponics—A Standard Methodology for Plant Biological Researches. Rijeka, Croatia: InTechOpen; 2012. pp. 39-66

[9] Gibeaut DM, Hulett J, Cramer GR, Seemann JR. Maximal biomass of Arabidopsis thaliana using a simple, low-maintenance hydroponic method and favorable environmental conditions. Plant Physiology. 1997;**115**(2):317-319

[10] Kratky BA. Three Non-Circulating Hydroponic Methods for Growing Lettuce. Acta Horticulturae; 2009;**843**:65-72

[11] Corbesier L, Gadisseur I, Silvestre G, Jacqmard A, Bernier G. Design in Arabidopsis thaliana of a synchronous system of floral induction by one long day. The Plant Journal. 1996;**9**(6):947-952

[12] Arteca RN, Arteca JM. A novel method for growing Arabidopsis thaliana plants hydroponically. Physiologia Plantarum. 2000;**108**(2):188-193

[13] Toda T, Koyama H, Hara T. A simple hydroponic culture method for the development of a highly viable root system in Arabidopsis thaliana. Bioscience, Biotechnology, and Biochemistry. 1999;**63**(1):210-212

[14] Arnon DI, Hoagland DR. A comparison of water culture and soil as media for crop production. Science. 1939;**89**(2318):512-514

[15] Hoagland DR, Arnon DI. California agriculture Experiment Station bulletin. Circular. 1950;**347**:32

[16] Huttner D, B-Z D. An improved, simple, hydroponic method for growing Arabidopsis thaliana. Plant Molecular Biology Reporter. 2003;**21**:59-63

[17] Jensen MH. Hydroponics worldwide. Acta Horticulturae. 1999;**481**:719-730

[18] Lippert F. Amounts of organic constituents in tomato cultivated in open and closed hydroponic systems. Acta Horticulturae. 1993;**339**:113-124

[19] Son JE, Kim HJ, Ahn TI. Chapter 17 - hydroponic systems. In: Kozai T, Niu G, Takagaki M, editors. Plant Factory. San Diego: Academic Press; 2016. pp. 213-221

[20] Dorneanu SA, Coman V, Popescu IC, Fabry P. Computer-controlled system for ISEs automatic calibration. Sensors and Actuators B. 2005;**105**:521-531

[21] Gutierrez M, Alegret S, Caceres R, Casadesus J, Marfa O, del Valle M. Application of a potentiometric electronic tongue to fertigation strategy in greenhouse cultivation. Computers and Electronics in Agriculture. 2007;**57**:12-22

[22] Kim HJ, Kim W-K, Roh M-Y, Kang C-I, Park J-M, Sudduth KA. Automated sensing of hydroponic macronutrients using a computer-controlled system with an array of ion-selective electrodes. Computers and Electronics in Agriculture. 2013;**93**:46-54

[23] Kraehmer H, van Almsick A, Beffa R, Dietrich H, Eckes P, Hacker E, et al. Herbicides as weed control agents: State of the art: II. Recent achievements. Plant Physiology. 2014;**166**(3):1132-1148

[24] Kraehmer H, Laber B, Rosinger C, Schulz A. Herbicides as weed control agents: State of the art: I. weed control research and safener technology: The path to modern agriculture. Plant Physiology. 2014;**166**(3):1119-1131

[25] Massonnet C, Vile D, Fabre J, Hannah MA, Caldana C, Lisec J, et al. Probing the reproducibility of leaf growth and molecular phenotypes: A comparison of three Arabidopsis accessions cultivated in ten laboratories. Plant Physiology. 2010;**152**(4):2142-2157

[26] Yan G, Liu H, Wang H, Lu Z, Wang Y, Mullan D, et al. Accelerated generation of Selfed pure line plants for gene identification and crop breeding. Frontiers in Plant Science. 2017;**8**(1786)

[27] Bulgarelli D, Schlaeppi K, Spaepen S, Ver Loren van Themaat E, Schulze-Lefert P. Structure and functions of the bacterial microbiota of plants. Annual Review of Plant Biology. 2013;**64**:807-838

[28] Harris SL, Pelaez CA, Shank EA. Monitoring bacterial colonization and maintenance on Arabidopsis thaliana roots in a floating hydroponic system. Journal of Visualized Experiments. 2019;**147**:e59517. DOI: 10.3791/59517

[29] Vandenbrink JP, Kiss JZ. Space, the final frontier: A critical review of recent experiments performed in microgravity. Plant Science. 2016;**243**:115-119

[30] Wolff SA, Coelho LH, Karoliussen I, Jost AI. Effects of the extraterrestrial environment on plants: Recommendations for future space experiments for the MELiSSA higher plant compartment. Lifestyles. 2014;**4**(2):189-204

[31] Barnwell P. Lighting the way to Mars: Plant growth for astronaut consumption. NASA Online Archive. 2017

[32] Paradiso R, De Micco V, Buonomo R, Aronne G, Barbieri G, De Pascale S. Soilless cultivation of soybean for bioregenerative life-support systems: A literature review and the experience of the MELiSSA project – Food

characterisation phase I. Plant Biology. 2014;**16**(Suppl 1):69-78

[33] Kuroda M, Ikenaga S. Single-tube hydroponics as a novel idea for small-scale production of crop seed in a plant incubator. Bioscience, Biotechnology, and Biochemistry. 2015;**79**(1):63-67

[34] Miller DM. Errors in the measurement of root pressure and exudation volume flow rate caused by damage during the transfer of unsupported roots between solutions. Plant Physiology. 1987;**85**(1):164-166

[35] Munns R, James RA. Screening methods for salinity tolerance: A case study with tetraploid wheat. Plant and Soil. 2003;**253**:201-218

[36] Gambrel D, Rink W, Mason T, Kottkamp J, Varjabedian A, Ross J, et al. Optimizing greenhouse corn production: What is the best root medium? Purdue Methods for Corn Growth. 2010;**17**:1-10

[37] Kottkamp J, Varjabedian A, Ross J, Eddy R, Hahn DT. Optimizing greenhouse corn production: What is the best fertilizer formulation and strength? Purdue Methods for Corn Growth. 2010;**14**:1-5

[38] Watson A, Ghosh S, Williams MJ, Cuddy WS, Simmonds J, Rey MD, et al. Speed breeding is a powerful tool to accelerate crop research and breeding. Nature Plants. 2018;**4**(1):23-29

[39] Ghosh S, Watson A, Gonzalez-Navarro OE, Ramirez-Gonzalez RH, Yanes L, Mendoza-Suárez M, et al. Speed breeding in growth chambers and glasshouses for crop breeding and model plant research. Nature Protocols. 2018;**13**(12):2944-2963

[40] Ochatt SJ, Sangwan RS. In vitro shortening of generation time in Arabidopsis thaliana. Plant Cell Tissue and Organ Culture. 2008;**93**:133-137

[41] Yao Y, Zhang P, Liu H, Lu Z, Yan G. A fully in vitro protocol towards large scale production of recombinant inbred lines in wheat (*Triticum aestivum* L.). Plant Cell Tissue and Organ Culture. 2017;**128**:655-661

[42] Zheng Z, Wang HB, Chen GD, Yan G, Liu CJ. A procedure allowing up to eight generations of wheat and nine generations of barley per annum. Euphytica. 2013;**191**:311-316

[43] Nagatoshi Y, Fujita Y. Accelerating soybean breeding in a CO2-supplemented growth chamber. Plant & Cell Physiology. 2019;**60**(1):77-84

[44] Paradiso R, Buonomo R, Dixon MA, Barbieri G, De Pascale S. Effect of bacterial root symbiosis and urea as source of nitrogen on performance of soybean plants grown hydroponically for bioregenerative life support systems (BLSSs). Frontiers in Plant Science. 2015;**6**:888

[45] Paradiso R, Arena C, De Micco V, Giordano M, Aronne G, De Pascale S. Changes in leaf anatomical traits enhanced photosynthetic activity of soybean grown in hydroponics with plant growth-promoting microorganisms. Frontiers in Plant Science. 2017;**8**:674

[46] Sheridan C, Depuydt P, De Ro M, Petit C, Van Gysegem E, Delaere P, et al. Microbial community dynamics and response to plant growth-promoting microorganisms in the Rhizosphere of four common food crops cultivated in hydroponics. Microbial Ecology. 2017;**73**(2):378-393

[47] Zhang Y, Butelli E, Alseekh S, Tohge T, Rallapalli G, Luo J, et al. Multi-level engineering facilitates the production of phenylpropanoid compounds in tomato. Nature Communications. 2015;**6**:8635

Chapter 3

Automation and Robotics Used in Hydroponic System

Alejandro Isabel Luna Maldonado,
Julia Mariana Márquez Reyes, Héctor Flores Breceda,
Humberto Rodríguez Fuentes, Juan Antonio Vidales Contreras
and Urbano Luna Maldonado

Abstract

Hydroponic system requires periodic labor, a systematic approach, repetitive motion and a structured environment. Automation, robotics and IoT have allowed farmers to monitoring all the variables in plant, root zone and environment under hydroponics. This research introduces findings in design with real time operating systems based on microcontrollers; pH fuzzy logic control system for nutrient solution in embed and flow hydroponic culture; hydroponic system in combination with automated drip irrigation; expert system-based automation system; automated hydroponics nutrition plants systems; hydroponic management and monitoring system for an intelligent hydroponic system using internet of things and web technology; neural network-based fault detection in hydroponics; additional technologies implemented in hydroponic systems and robotics in hydroponic systems. The above advances will improve the efficiency of hydroponics to increase the quality and quantity of the produce and pose an opportunity for the growth of the hydroponics market in near future.

Keywords: hydroponic systems, sensors, microcontrollers, automation, neuronal networks, robotics

1. Introduction

It is estimated that the total world population could reach 9.15 billion in 2050 [1], and to increase the global food production, even more advances in agriculture must be made intensive in crop yields and in practices that are more friendly with the environment. Hydroponics is a method of growing plants in a water solution without soil. If the roots are suspended in a liquid medium or supported using an inert medium, the system is known as Nutrient Film Technique (NFT). In NFT systems, the plants (lettuce, leafy crops and herbs) are grown in channels (gullies) and fed continuously at a rate of approximately 1 L min^{-1} (**Figure 1**).

On the other hand, if the roots are floating (pool), the system is known as deep floating technique (DFT). The DFT systems (**Figure 2**) are long, cement or wood rectangular reservoirs and lined with a durable polyliner. To keep the plants in net pots, holes are perforated in a foam board which rest on the surface of the water.

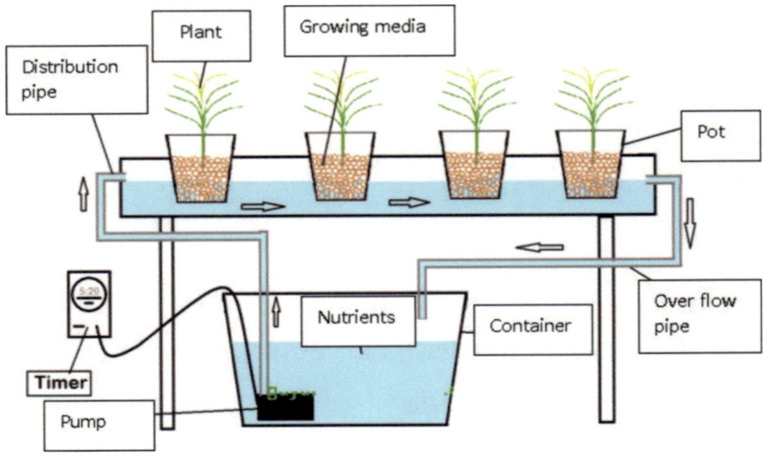

Figure 1.
Nutrient film technique.

Figure 2.
Deep floating technique (courtesy of Hydroponicsfarm).

Aeroponic systems are very similar to NFT systems, differing primarily in the spatial arrangement of cultivation channels. The cultivation space is optimized for the aeroponic plants are grown suspended in air, having as support PVC pipes which can be arranged horizontally or vertically, enabling a better exploitation of areas and installing a larger number of plants per square meter surface of the oven, obtaining thus a direct increase of productivity [2, 3]. Hydroponic systems, such as the deep flow technique, nutrient film technique or aeroponic systems, are essential tools in plant factories [4]. To accomplish with this, hydroponic systems must collect a lot of information, since this allows a better diagnosis of the problems and better understand the development of hydroponic crops. Automatic sensors not only have the ones that can be read at predefined intervals, but also the readings of these sensors are stored so that higher results can be obtained for analysis and diagnosis resulting in higher crop yields and friendlier practices with the environment. These days, there are microcontrollers (**Figure 3**) on the market that are compatible with a wide variety of sensors and can be used for automatic monitoring and robotics.

The emergence of Internet of Things (IoT) has allowed farmers to automate the hydroponic culture (**Figure 4**). Monitoring of water level, pH, temperature, flow and light intensity can be regulated using IoT, which allows for machine to machine interaction and controlling the hydroponic system autonomously and intelligently employing deep neural networks [5]. The pH of the nutrient solution for most nutrient film technique is 6.0–7.0 for most plants grown in recirculating nutrient solution and 5.4–6.0 for substrate culture [6]. There are also powerful computers that could store all this information and build a big database.

Figure 3.
Huertomato microcontroller for measuring of humidity, water and air temperature, light, pH, electrical conductivity (courtesy of Arduino).

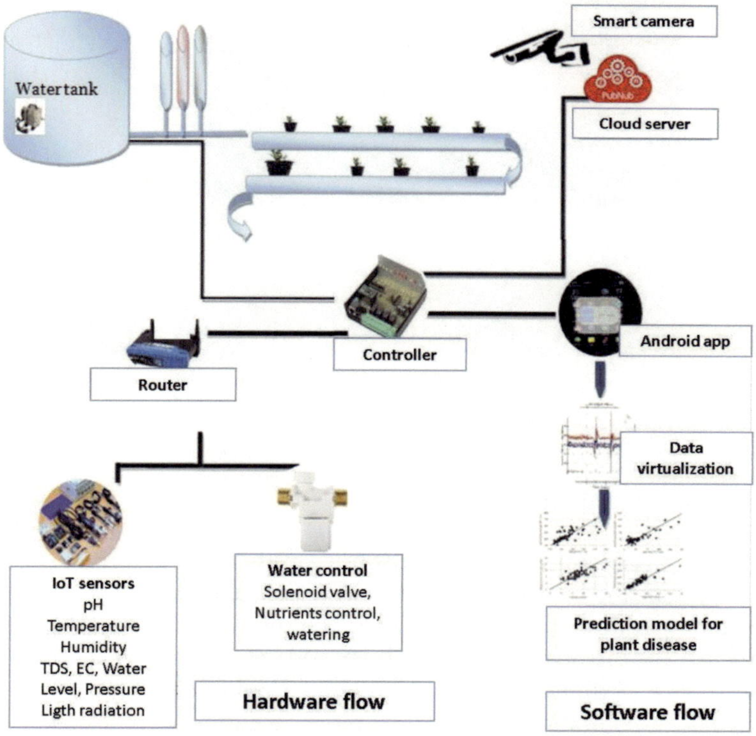

Figure 4.
Schematics of internet of hydroponics (courtesy National Institute of technology, Trichy, India).

It has been implemented a smart hydroponics system (**Figure 5**) that automates the growing process of the crops using Bayesian network model, which classifies and predicts the optimum value in each actuator to autonomously control the hydroponics farm [7]. Finally, we raise topics related to robotics for hydroponic systems (**Figure 6**). Hydroponic systems cover approximately 35,000 ha in the world and further research is needed to develop new hydroponic systems to reduce the cost of energy and materials required for crop production. Therefore, this chapter aims to be a practical guide to those interested in hydroponics automation and robotics to produce vegetables.

Figure 5.
Automatic grow cabinets for growing plants at home (courtesy of HG-hydroponics).

Figure 6.
Robot for hydroponic systems (courtesy of Iron Ox Company).

2. Automation in hydroponic systems

All possible variables in root zone must be monitored for automation of the hydroponic system and sensors of pH, the electrical conductivity (EC), light, the ambient temperature, the temperature of the solution, the humidity and the carbon dioxide, the dissolved oxygen and the oxidation–reduction potential must be considered as they directly affect the growth of hydroponically grown plants (**Figure 7**). The transpiration can be measured with either water ultrasound level sensors or load cells. If the area or volume of culture is large, several sensors must be placed to adequately control the entire crop. Ion sensors (17 essential elements in plant nutrition) are still studied for their durability and stability [8].

2.1 Hydroponic system design with real time OS based on microcontroller

It was developed a complete automation hydroponic system for maintaining stable electrical conductivity, pH, growth light and monitoring CO_2, temperature and humidity. The system consisted of an ARM Cortex-M4 microcontroller running ARM (**Figure 8**) embedded operating system, the official real time operating system (RTOS). The system read the pH level and nutrients on the nutrient solution of hydroponics system, as well as the temperature, humidity, CO_2 levels and the

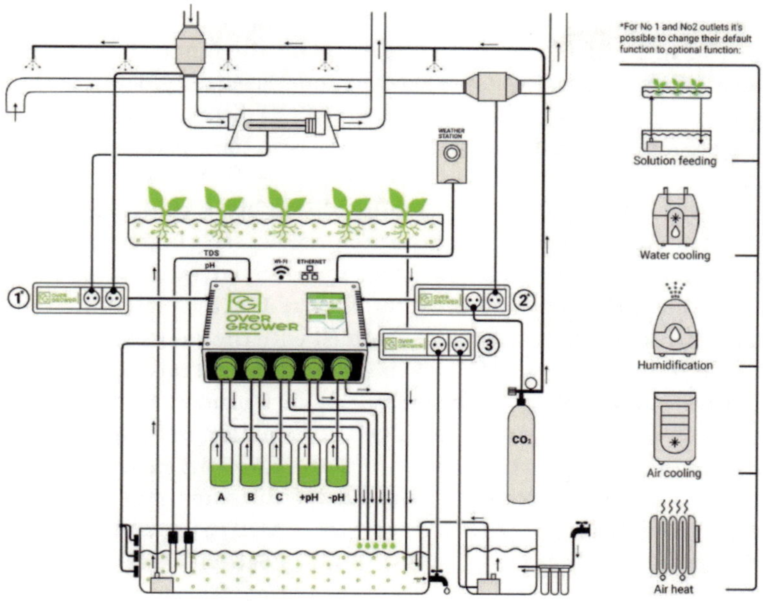

Figure 7.
Hydroponics automation system (courtesy of over grower).

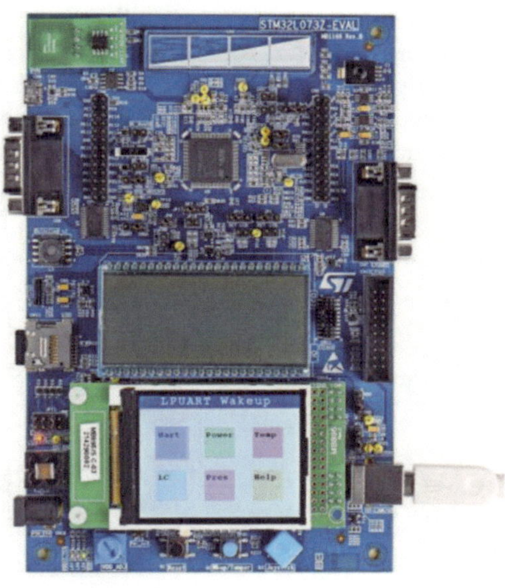

Figure 8.
ARM cortex-M4 microcontroller (courtesy of developer arm).

light intensity around the system; in addition there were LED light three lines each a different configuration on each line that was used for the lighting of plants and light color was selected and the system data were saved in SD Card.

In addition, the system was capable to control desired concentration level with variation of less than 3%, pH sensor showed good accuracy 5.83% from pH value 3.23–10. Growing light intensity measurement was 105 μmol/m^2/s therefore, the lights were turned on at least 17 h/day to fulfill plant light requirement. RTOS gave good performance with latency and jitter less than 15 μs, system overall show good performance and accuracy for automating hydroponic plant in vegetative phase of growth. If the system was turned on, the computer program turned off three pumps (stir pump, water pump and the dosing pump). After initialized an LCD module, then initialized serial and serial to PC for CO_2 sensor. The program read the system configuration data that were stored in the SD card and initialized global variables with the configuration. Later program will update the LCD display. The program started up the sensor to read data sensors, push button to read the buttons provided, mixing to perform compounding nutrients hydroponics, timer flush to set watering plants, timer lights to regulate time lighting plants by LED lights. A timer was started up for minimum water and a sensor to detect the presence of water in nutrients within of a container. Once the water was activated, then timer watered the plants. The pH sensor recorded the initial pH value in the solution, then adding the pH solution up 5 mL and compared the pH sensor measurements obtained with the instrument (**Figure 9**).

A total of 600 s was taken by DHT22 humidity sensor sampling every 30 s and the readings were compared with the measuring instrument. Twenty-five minutes were taken by MH-Z19 sensor with readings every minute in rooms, results were compared with measuring instruments. The system initially provided nutrients for 5 mL and then the system recorded and calculated the amount of nutrients needed. A distance of 30 cm from LED to plant hole was settled to use a meter for Quantum PAR (photosynthetically active radiation). A LED coefficient was obtained by dividing average light intensity (ALI) with lux. The coefficient of LED and ALI can be used to find daily light integral (DLI) during 17 h. RTOS performance was obtained using square wave input signal and measuring input signal versus output signal delay using oscilloscope. The difference of humidity data retrieval between DHT22 sensors and measuring devices was very small. CO_2 data retrieval between MH-Z19 sensor and measuring devices at room had a difference for each room relatively equal amount, then for the sensor MH-Z19 in this case with a correction factor, so the results obtained are close to the results of measuring instruments. A correction factor of 260 ppm was used for the MH-Z19 sensor against the initial

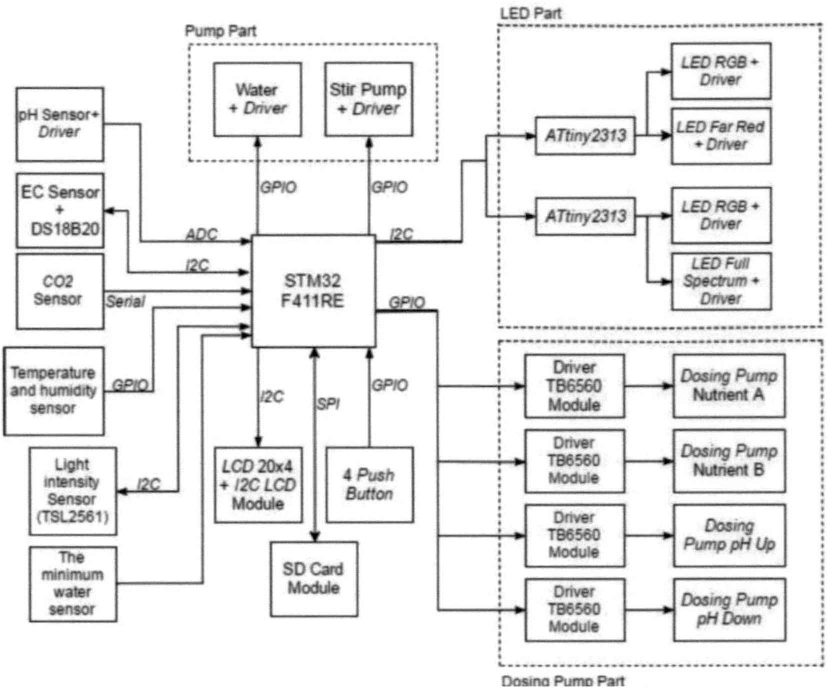

Figure 9.
Block diagram of system.

value. The growing light intensity was measured at 25 cm from light source. The result showed each growing light produces difference intensity ranging from 6.03 to 10.74 μmol/m^2/s, to fulfill plant light requirement on at least 17 h day^{-1}. From the results of pH meter sensor, the difference obtained is so small so that the pH meter sensor can be used to read the pH suitably. When taking data for RTOS experiment (**Figure 10**), the main programs still running while the experiment still ongoing.

The yellow signal is a given signal and the other signal is a signal output from each thread. Time latency (in microseconds) was very small. The results showed that hydroponic automated system performed well. RTOS ran all the tasks with a latency less than 15 μS. Environment sensor overall showed good result, temperature reading error was less than 4%, humidity reading less than 5.36% and CO_2 sensor accuracy was calibrated 260 ppm from initial value. System was capable to mix nutrients in 80 s with error less than 3.48%. Light intensity measurement showed different result for different color spectrum in order to fulfill daily light plant requirement we need to turn on the light at least 17 h day^{-1}. The vegetable grew well and can be harvested in 5 weeks [9].

2.2 pH fuzzy logic control system for nutrient solution in embedded and flow hydroponic culture

The fuzzy-based control system was developed for maintaining a proper acidity level of nutrient solution used in potted flower cultivation of Chrysanthemum embedded and flow hydroponic cultures. Two control valves maintained the nutrient solution pH at a desired set point as follows: (1) acid valve (to manage the addition of acid solution necessary) and (2) base valve (to keep the addition of base solution necessary) (**Figure 11**). The developed control algorithm was based on membership functions of fuzzy arrangement.

Fuzzy rules had 21 linguistic statements to achieve smoothness, by trials and errors using the membership functions based on the operator skills and experience. The fuzzy logic controlled nutrient solution pH and increased the smoothness of the pH the during control course. The culture vessel consisted of six blocks, each of which containing four potted flowers. The nutrient solution flows into and fills the cultivation bench until a certain level, 5–10 cm from pot base. The embedded system kept the plant growth media in 10 min, before it then flows back into the tank and flows into the next block. The flow rate of the nutrition used in this experiment was 2.4 L min^{-1} and the measuring apparatus was Hanna pH-meter (HI8710E model).

The control system maintained 0.3 M H_3PO_4 and 0.4 M KOH, which flowed constantly from Marriott tube. The valve used was of solenoid type with 1/8 in. in diameter. Calibration of the pH-meter was done on voltage basis using PCL-812PG

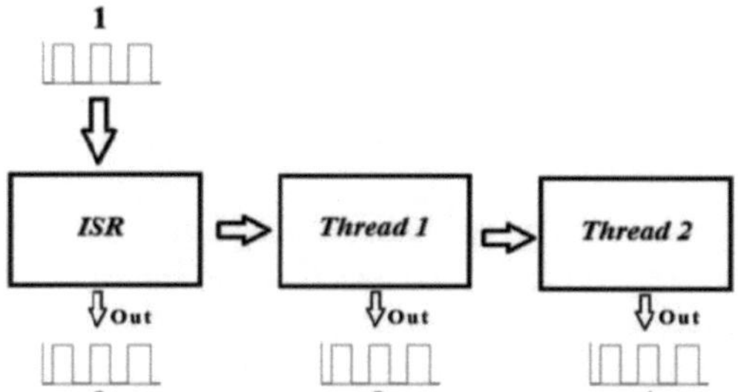

Figure 10.
Experiment of RTOS.

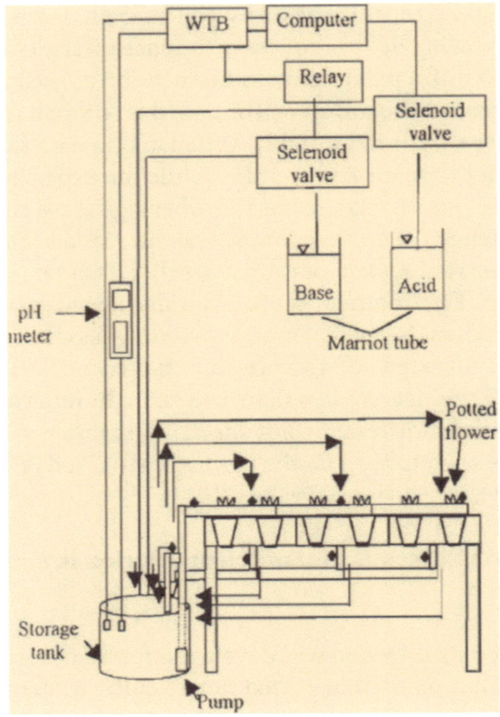

Figure 11.
Embedded and flow system with pH control system.

interface. Marriott tube was also used to calibrate the flow rate as well as on the relay circuit. The measurement result of the pH of nutrient solution was in the form of DC voltage and was transferred to 88 shunt circuit in order to get input voltage at a range of 0–5 V conforming to the working voltage of the PCL-812PG interface. This voltage became the reference digital signal for the computer to conduct data processing with control program. The output of the control action was the duration of the solenoid valve opening depended upon the input signal. A solenoid valve was activated by a relay circuit, which obtained voltage from the computer.

Process error (E) was calculated based on the difference between the set point (Sp) and the actual pH. If an E positive value was obtained, it indicated that the position of the actual pH was above the Sp and negative value of E indicated that the position of the actual was under the Sp. The error difference (dE) was the change in E to time. The error difference (dE) was the change in E to time. If the dE were positive, the error E had the tendency to increase. Conversely, if dE were negative, the error E decreased. Every numeric variable was plotted into a fuzzy system consisted of Large Positive (LP), Fair Positive (FP) and Small Positive (SP), Zero (ZO), Large Negative (LN), Fair Negative (FN) and Small Negative (SN). The control action was based on decision matrix in which there are criteria of Quick Acid (QA), Fair Acid (FA), Slow Acid (SA), Neutral (ZO), Quick Base (QB), Fair Base (FB) and Slow Base (SB).

The measurement result of the pH of nutrient solution was in the analog form of DC voltage and was transferred to defuzzification by means of weighting to the absolute membership value from every label with the membership degree obtained. A change in valve opening time, either for base tube or acid tube was due to the final output of the fuzzy. The computer program for the control system was developed using the Pascal language in DOS environment (**Figure 12**). The output voltage from the PCL-812PG had a range of 0−+5 V. The debit of the base and acid

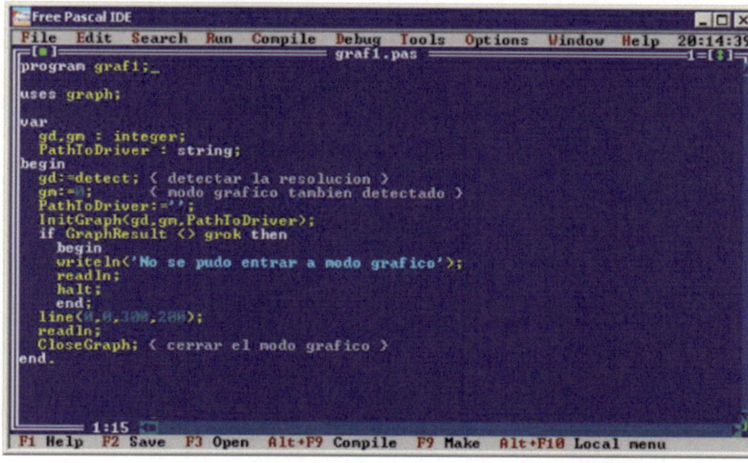

Figure 12.
Pascal environment.

flows from the Marriott tube was kept constant at 1.3 eels for base solution and 4.3 $cm^3\ s^{-1}$ for acid solution. There were differences in the heads of the air inlet and outlet at the Marriott tubes for base and acid solutions, respectively. The initial pH of the solution was above the set point and kept on moving to reach the pH = 6. To change the pH solution from 7.0 to 6.0, it took 26 s and a 100 s to increase pH from 6.0 to 7.0. That indicated that at the same period of time $[H^+]$ freed by the H_3P0_4 acid was more than that of the ion $[OH^-]$ freed by the KOH base.

The supplied voltage from PCL to relay circuit was on when the voltage reached 1.4 V and off when the voltage decreased to 1.1 V. The pH of the nutrient solution in first block can be controlled to approach the set point of pH = 6. To decrease the pH toward the set point it requires 68 s. After reaching the set point. The pH of the solution did not change very much due to the small change in $[H^+]$ concentration. Moreover, the straight line approaching the set point tendency of the error curve during the control indicates that the fuzzy logic control can maintain the solution pH at the set point. An overshoot not occurred in this pH control. The nutrient solution pH in second and third block can be controlled faster than in first block. The same phenomena occur in third block and the following blocks. In that manner, the set point indicates that the fuzzy logic control can maintain the solution pH at the set point. Both of valves frequently open in turns since the control load was still high at the start up. This frequency decreased at the following blocks [10].

2.3 Hydroponics in combination with an automated drip irrigation system

A grapevine rootstock in hydroponics in combination with an automated drip irrigation system was developed, which consisted of a hardware and software of the automated hydroponics system for grapevine in pots. Each pot had the same amount of fertilizer and the drip irrigation system was used. It was also constructed a time-based closed loop hydroponics and used a microcontroller for supplying the water to the pots (**Figure 13**). The irrigation system consisted of a 200 L water storage tank, containing Hoagland solution, which was modified to regulate the optimum 6.2 pH and electrical conductivity levels between 1.0 and 1.3 dS^{-1} for the green cuttings of grapevine. Nutrient solution has been renewed when its EC level reached to 1.5–1.8 dS^{-1}. A submersible pump operating at 12 DCV was installed inside the water storage tank. A steel structure was built to keep pots at height of 1.5 m from the ground level of the greenhouse and water storage tank and controller

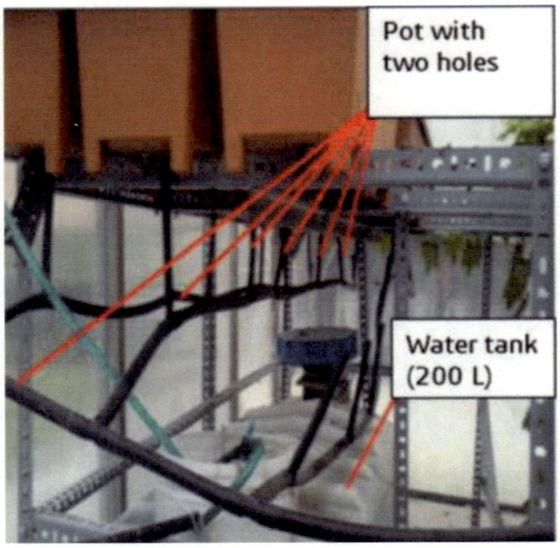

Figure 13.
Grapevine experimental setup.

circuit were installed immediately under the pots. The excess water was easiest to return to the reservoir by the drainage pipes connected to the drainage holes of the pots. Electrical conductivity of the irrigation water (ECw) was measured by an EC59 meter. Pots were irrigated with the same amount of nutrient solution. The required water was supplied by using 16 inches of diameter pipes with 4 L h^{-1}-drippers at a spacing of 33 cm, with three drippers serving each pot. Some connection apparatus and valves were used in the irrigation system to integrate all items.

At the beginning of the test, all substrates were filled up to field capacity, then the automated system started irrigation at 4 h intervals and run the submersible pump only 1 min throughout the whole growing season so that this irrigation management kept the soil moisture at the level of field capacity in each substrate since excess water was drained to the reservoir back after each irrigation event. The controller circuit, in which main power supply was 12 DCV, providing power to the controller and relays, but it was reduced to 5 DCV for microcontroller by using a regulator of 7805 and relay (**Figure 14**).

The program providing the automation in the hydroponics system was simple and basic and very easy to load into the memory of the microcontroller, which repeated the actions throughout the whole growing season. The dosage of water was determined according to the pumping time of water. The microcontroller switched on relaying to pumping water to the root territory only for 1 min. After that, supplying of water has been stopped to the pump and then waited for 4 h of interval for the next irrigation session. The system took over the irrigation events successfully for the whole growing season. The system conveys a properly balanced nutrient solution to the plant root area. The system saved water and fertilizer, but the water level in the reservoir must be checked with 2- or 3-weeks interval or water level sensor should be added to the controller circuit. Perlite due to its characteristics has more advantages as being used in the hydroponics system as compared with peat and peat + perlite (1:1, v:v). This system can be used for small producers from small hydroponic systems [11].

2.4 Automatic system for hydroponics (HydroAS)

HydroAS could produce fodder in 6 days. The system controlled autonomously the desired agronomic conditions for production and fodder flow. The automatic

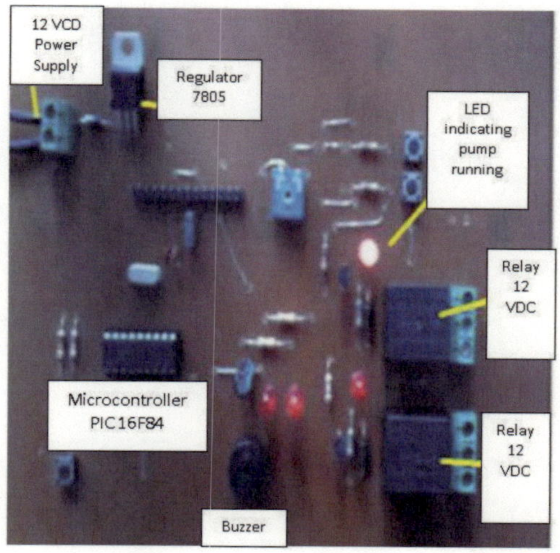

Figure 14.
Controller circuit.

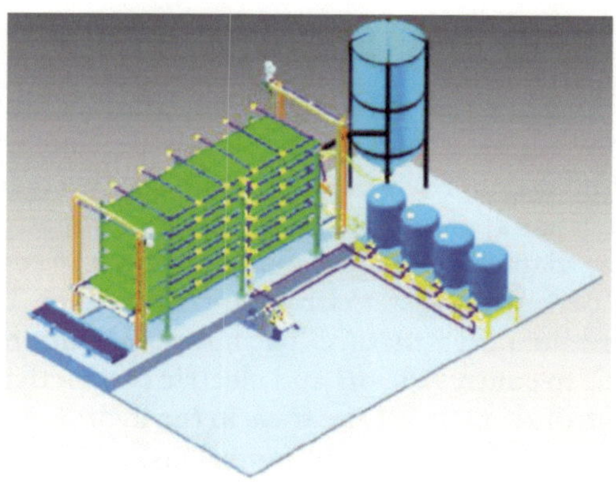

Figure 15.
Fodder production system.

solution comprised: the mechanical structure, the mechanical and hydraulic components, and the control system to automate the hydroponic automatic system. The mechanical structure consists of the following parts: (a) mechanical structure of six storeys; (b) conveyor to exit the produced fodder off the system; (c) two elevators, at each top of the six storey structure; (d) a fodder sowing system, which placed the seeds in the trays; (e) two pushers at each of the elevator, which pushed the trays in the structure and (f) unloading system, which extracted the finished fodder in the trays (**Figure 15**).

The electrical components comprised the power, sensor and actuators circuits (**Figure 16**). The power circuit consisted of depicted the protections and transformers to obtain 24 DCV to supply the S7–300 programmable logic controller (PLC), sensors and the command circuit, which were mainly digital, inductive and magnetic, that indicate start/end limits for the actuator's movement such as level sensors applied in the nutrients and water reservoirs. The actuators used were (a) two motors for the vertical movement of the elevators; (b) two pneumatic cylinder for the pusher; (c) two worm motors for the sowing platform and (d) two gearmotors for the rotational joint of the sowing and unloading platforms. The hydraulic system supplied the nutrients and the flow of water in the system and consisted of

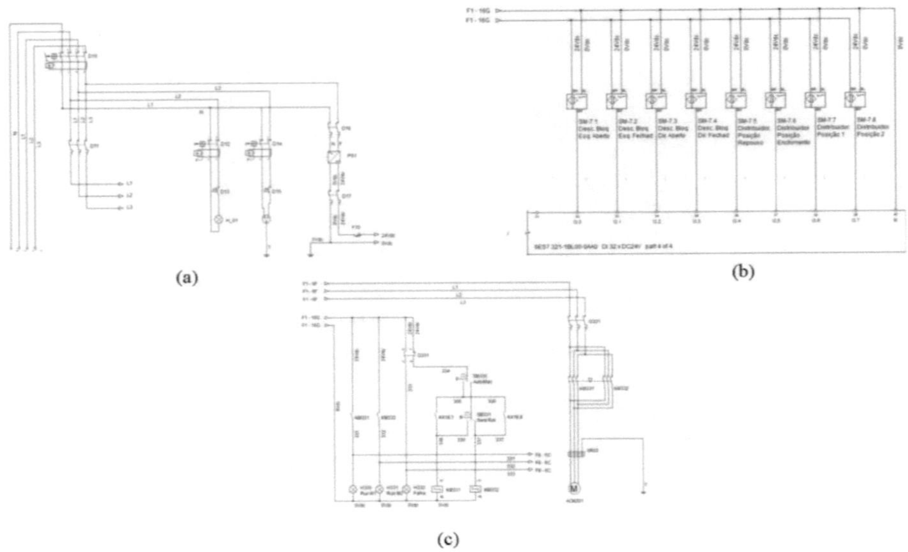

Figure 16.
Power circuit (a), sensor circuit for the unloading part (b) and actuator circuit for the elevator one (c).

two irrigation pumps to generate redundancy. The hydraulic circuit comprised the valves, nutrients and water reservoirs, the six storeys pipeline with the irrigation micro-jets, and the 2 m^3 water return reservoir with the two redundant pumps. The chosen microspray jets operate at 1 bar and have the capacity of 1 L h^{-1}, with 0.8 m maximum spray diameter area. The pipeline structure consisted of six storeys and watering pumps performed system irrigation three times. The system controls the actuators of the mechanical structure. The nutrient solution control is also performed in the PLC, to control the pH and electric conductivity, while mixing the nutrients and the control sequence of the trays in the hydroponic system.

This sequence definition is a high-level control task, while the low-level actuator control is performed in inner loops and is programmed directly on the motor drives. The system starts when the first storey is filled with trays. When the seeds are placed in the tray, in the first storey near a first elevator, the next step is to elevate the tray and push it to the structure in the second storey. On the other side of the structure, a second elevator, receives a tray that was pushed as consequence of the previous movement. This tray is then elevated to the next storey. This process is repeated until the tray reaches the end. When this happens, the elevator number two descends to the first storey and unloads the produced fodder to the conveyor. After unloading this tray, it is pushed to the first level for washing, and the next 6 days cycle then starts, to produce new trays full of fodder. This system was simulated in Matlab SimMechanics, showing its proper operation for the mechanical and electrical parts. The development phase of the fodder was tested, and validated, which benefits the agricultural holding [12].

2.5 Expert system-based automation system (HES)

HES was developed to minimize the labor force used in the process of hydroponics, the total amount of time spent in agricultural process, human-based errors, as well as, the control of hydroponics greenhouse plant production. All these processes are conducted by a computer unit where the relevant programs are loaded. The system defined the values that belong to the input parameters by using the output parameters that are used by the user (**Figure 17**). The input

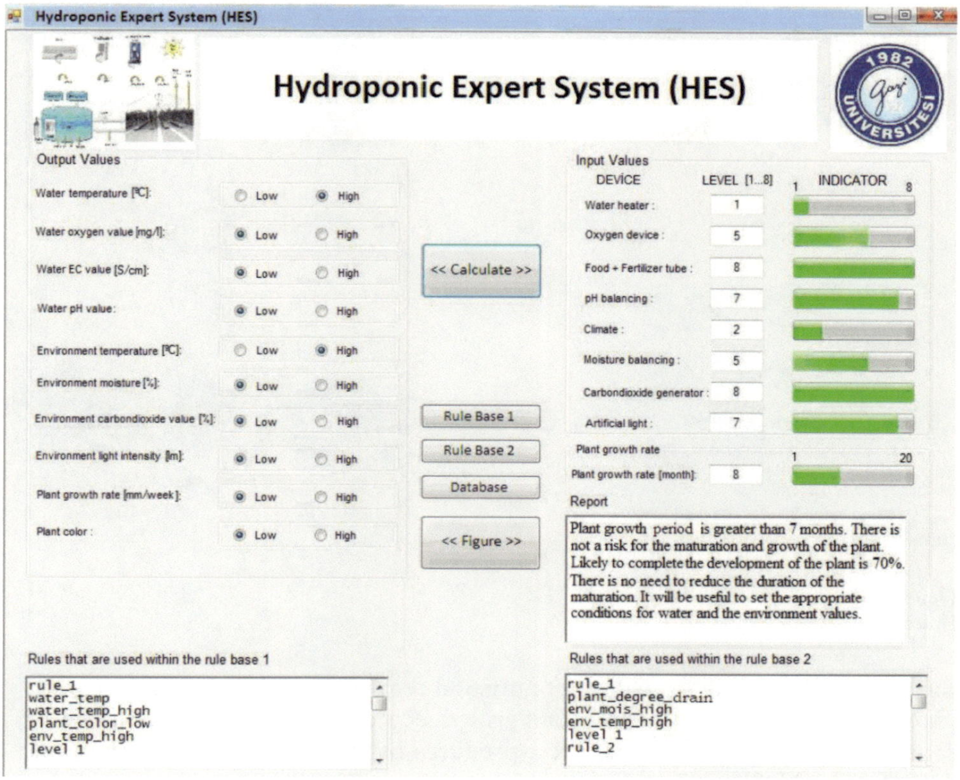

Figure 17.
Graphic user interface of HES.

parameters prepared the optimum growth environment for the plants. User interface controlled the knowledge base of the expert system and entered data and realize operations. All parameters were taken in to consideration in order to create controlled environment exactly. Knowledge base is continuously in a process of improvement and human experts would add new knowledge to knowledge base or modify the existing knowledge heuristics when new situations occurs. The data base was made up of real conditions that summarize the current situation of the problem and quality-value pairs. By all these output parameters level management, the total level grade can be attained, and this can determine the development period of the plant.

This system had two rule bases. The first one was the rule base that constitutes the input parameters and the second one was the rule base that determines the growth period of the plant. The growth period of the plant is determined by adding the values of plant drain degree, plant nutrition degree, plant deterioration degree, plant photosynthesis degree and plant growth degree. The inference engine, had the function to produce the results that the system needs by using the data in the knowledge base and by interpreting the rules of the system as follows; the user interface of the HES software and the relevant output values taken from the greenhouse system and the sensors are interpreted and translated into linguistic expressions such as low-high. When the system finds a rule that matches the related values in the rule base, it attributes this level value as the level value. It is used for all parameters temperatures, oxygen device, nutrition and the operating levels for water heater, fertilizer tube, pH balance, conditioner, moisture balance, carbon dioxide producer and artificial light are attuned, creating the appropriate conditions for the greenhouse system (**Figure 18**). Plant growth period is attained from

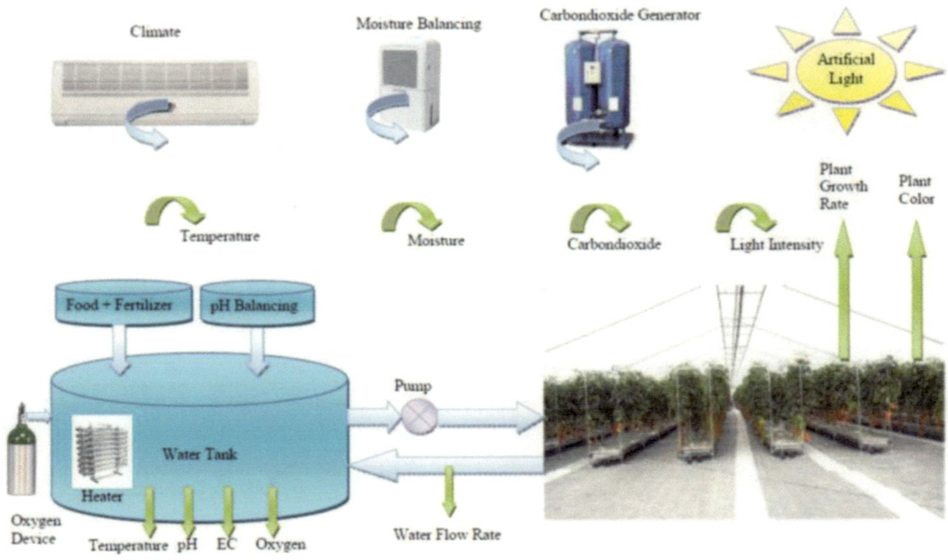

Figure 18.
Hydroponic system setup.

the total of plant drain degree, plant nutrition degree, plant deterioration degree, plant photosynthesis degree and plant growth degree parameters. After all these processes are completed, reports were produced by the system, based on the plant growth period [13].

2.6 Automated hydroponics nutrition plant systems (AHNPS)

AHNPS was placed in a special chamber or vessel and the nutrients were supplied directly to the hydroponic roots at any given time (**Figure 19**). Microcontroller (Arduino Uno) will control the flow of nutrient solution on the vessel automatically, and the microcontroller can be controlled from Android smartphone. This system had an embedded program module. The microcontroller worked in real time to setup the alarms on nutrient pumps. If alarm is enabled, a relay will be also activated, and then the pump will drain the nutrition solution on the plant. If alarm is deactivated, the relay will be turned off and the pump will stop supplying. Moreover, it has been designed a virtuino application on Android smartphone that serves to check the water level and temperature around the plants. Before starting the design of a virtuino application, it first provides the data storage using the features of thingspeak.com.

The hydroponic flow system starts from the detection of a proximity sensor and a temperature sensor (**Figure 20**). The sensor will detect the water level in the hydroponic tube and the temperature sensor to detect the room temperature. The detection sensor was connected to a relay that in turn was attached to the microcontroller port. When the relay port pin is lower than the specified height, the water flow will be run on the water pump to irrigate the plant. If the relay of port pin is turned on, it means the water level is above of the specified height, and then the water pump will stop being water, in that manner the water flow was regular. Time and date were displayed at any time of the process in an LCD screen. The pumps are used not only to increase water but also to add nutrients to the hydroponic tube. The water pumps were used for water recirculation and relays used to control both nutrient solution as well drain pumps. The system mechanism worked as follows: HC-SR04 ultrasonic sensor detected the height value of

Figure 19.
Experimental hydroponic system configuration.

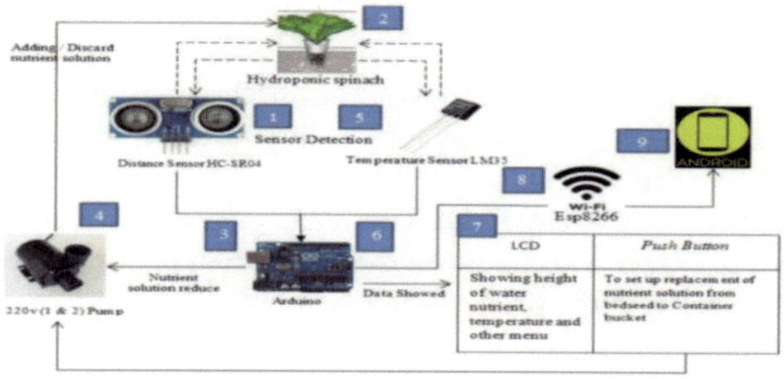

Figure 20.
Automated hydroponics nutrition plant system.

nutrient solution in hydroponic plants by the parameter of the high of water (in cm) unit and the temperature. The LM-35 sensor detected the temperature in Celsius degrees.

The ultrasonic sensors measured the distance of water based on ultrasonic wave. The difference between the transmission time and the reception time became the water distance. The system started working by using of sensors connected to the electrode. The output of this electrode will be forwarded to the microcontroller as an entry point to be processed by the microcontroller. The microcontroller received this voltage signal and compared it with the previous value and decided based on that input signal. Based on this voltage the microcontroller decided whether to drain the water at the pump or not. All the commands on microcontroller (Arduino Uno) can be controlled from a smartphone-based Android. It was observed that this hydroponic plant grows well with proper water and nutrient usage because it is controlled by the microcontroller. The rate of hydroponic plant growth was faster when compared with plants with soil-grown systems. The WiFi module sent the water level information and the temperature values of the plants area. This value was compared to the value in LCD microcontroller and on Android smartphone application. This value is directly obtained from the sensor and sent to the Arduino. After the water level 5 cm in the nutrient tube then the pump stops, and water did not flow anymore. The average temperature for five tests was 28.43°C. The relationship of water height in the nutrient tube with time is recorded continuously by the ultrasonic sensor on several measurements. The maximum level of water was 6 cm in the hydroponic tube. The sensor detected if the water level decrease in hydroponic nutrition tube. If the water level has decreased, then the sensor will perceive what occurs and automatically the water pump will be turned on to increase the water level on the hydroponic nutrient tube [14].

2.7 Hydroponic management and monitoring system for an IOT-based NFT farm using web technology (Hommons)

It was created a hydroponic farm management system that could monitor water temperature, water level, higher densities of nutrient solution and the acidity of a nutrient solution using sensors are related and connected to the microcontroller via a website. Hommons used a 20 W solar system, which consisted of a solar cell panel, controllers, battery and DC to AC inverters. The ESP8266 module was used as a communication medium through a wireless network to the internet and integrated with objects that have connection to the internet. Systems can be accessed through the web page using browser based on the server address. The core material of the PVC pipe tool with 3 in. of diameter as his planting medium and ¾ in. of diameter PVC to flow the nutrient solution. The plastic box reservoir served to accommodate any mix of nutrient solution in water. Hommons hardware design relationship of the NFT consisted of sensors, actuator, microcontroller, ESP8266, wi-fi access point, microcomputer (Raspberry Pi) and power supply. In addition, some Raspberry Pi 2 microcomputers served to accommodate the webserver and brokers. Communication technologies on this system using 802.11 or better known as Wi-Fi by using the internet (**Figure 21**).

The power supply using voltage 5 DCV and 2A. Various environmental sensors had been installed to detect any change in the physical or chemical environments and sensors became the input to the process management of NFT. After the user successfully performs the login process so the system redirected the user to the main page heading. There were two buttons on the sidebar navigation. On the main content there were four columns that display data from the sensors-sensors on the NFT hydroponic farming tools, such as nutrient levels, nutrient pH levels, temperature, nutrient and nutrient EC and parts per million (ppm) level. In the navbar a notification function button displayed the alarm or warning to the user while the settings button function settled the system (restart and shutdown) and logout of the system. Automation settings pages were divided into two parts: first part with its own set of pH and ppm values are desirable way entering the value in the textbox. The second automation page contained a selection of plants type which pH and ppm value have been set before, so farmers only need to choose the type of plants that they maintained to grown. After the hardware and sensors on the hydroponic NFT management system were integrated, the sensors (ultrasonic sensors, pH, temperature and EC) needed to be tested to quantify the level of accuracy. The system testing used the

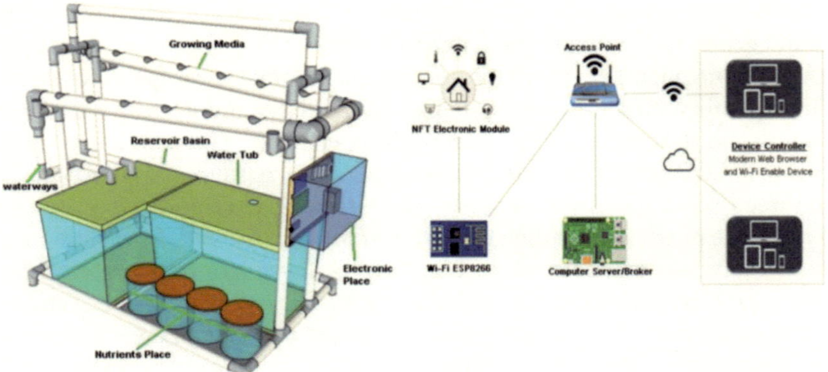

Figure 21.
Hommons hydroponic system.

original plant samples to find out if the plant is growing well. The plants used in this test are pokchoy, lettuce and kale at the teen age period (after nursery). Plant growth was observed by taking pictures of the plant for a few days [15].

2.8 Neural network-based fault detection in hydroponics

It was developed a fault detection model for hydroponic systems, with a feed-forward neural network. Mechanical, sensor and biological faults were considered: a preliminary detection system detected the existence of any faulty situations. Finally, the developed network, only considered two first kinds, mechanical and sensor faults. Biological faults, because of their particularities, were treated separately [16].

Other model based on a feedforward neural network predicted pH and EC changes in the root zone of *Lactuca sativa* cv. Vivaldi grown in a deep–trough hydroponic system. The neural net had inputs as follows: pH, EC, nutrient solution temperature, air temperature, relative humidity, light intensity, plant age, amount of added acid and amount of added base and two outputs: pH and EC. A combination of network architecture and training method was one hidden layer with nine hidden nodes, trained with the quasi–Newton backpropagation algorithm which was the most suitable and accurate (**Figure 22**). The model was capable of predicting pH at the next 20–min time step within 0.01 pH units and EC within 5 $\mu S\ cm^{-1}$. Simpler prediction methods, such as linear extrapolation and the lazy man prediction, value of the previous time step, gave comparable accuracy much of the time, though, they performed poorly in situations where the control actions of the system had been activated and resulted rapid changes in the predicted parameters. In those cases, the neural network model did not encounter any difficulties predicting the rapid changes. Thus, the developed model successfully identified dynamic processes in the root zone of the hydroponic system and accurately predicted one–step–ahead values of pH and EC [17].

2.9 IoT-based intelligent hydroponic system

An IoT-based intelligent hydroponic plant factory solution called PlantTalk was developed. PlantTalk intelligence could be built through an arbitrary smartphone. PlantTalk was flexibly to configure the connections of various plant sensors and actuators through a smartphone. Python programs for plant care intelligence through the smart phone were convenient (**Figure 23**). Automatic LED lighting, water spray, water pump and so on were included in the developed plant-care intelligence included and so on. For instance, it was showed that the PlantTalk

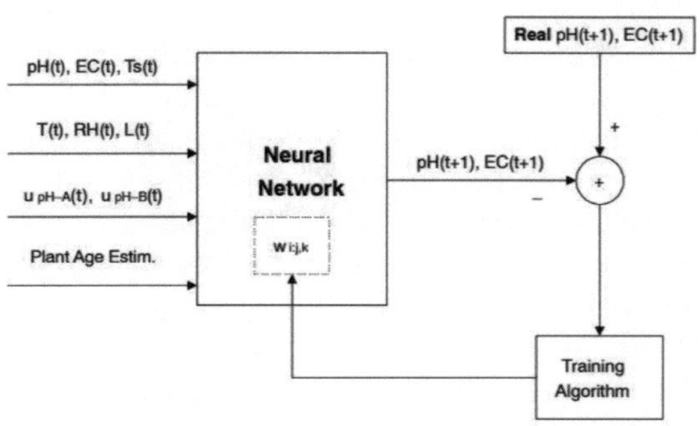

Figure 22.
Neural network model inputs and outputs and training process.

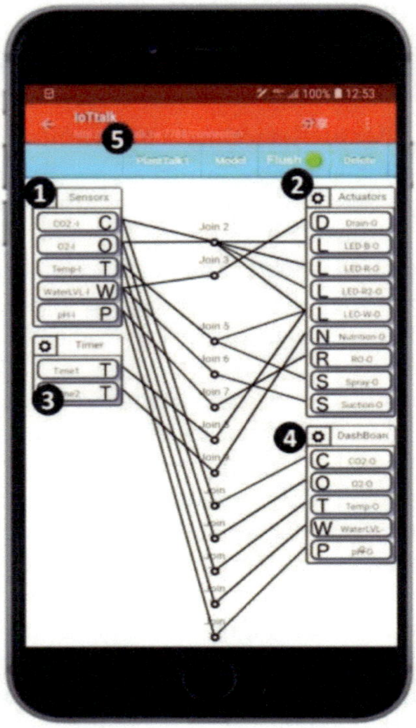

Figure 23.
PlantTalk in a smart phone.

intelligence effectively lowers the CO_2 concentration, and the reduction speed is 53% faster than a traditional plant system. AgriTalk for a plant factory is an extension of PlantTalk [18].

2.10 Other technologies used in hydroponic systems

It was applied ultrasound and dissolved oxygen supersaturation as external for controlling the growth rate of plants in hydroponics as well as maintaining the product quality. In the case of the leaf lettuce growth in hydroponics with exposure to 28-kHz ultrasound and dissolved oxygen supersaturation up to 36 mg L^{-1} at 20°C and peak-to-peak pressure at 20 kPa or larger worked as the growth inhibitor of the leaves and the roots; in addition, oxygen supersaturation became a growth promoter, without any degradation of chlorophyll in the leaves [19].

On the other hand, liquid separated reactor and a high voltage power supply based on a 20 kHz inverter neon-transformer were developed to archive the treatment with high energy efficiency, a low initial cost and a low running cost. The performance of the system on bacteria inactivation in the nutrient solution was evaluated in a continuous treatment system operation and the results showed that the standard plate count for background microflora and *R. solanacearum* is drastically reduced by the plasma treatment and is not detected after 8 days treatment. The nutrient solution was decontaminated by 4 log cycle with plasma treatment under the continuous operation condition [20].

Other researchers applied electro-degradation (ED) to the culture solution in order to degrade their root exudates and improve growth, yield and quality of strawberry. They used four types of nutrient viz. renewed, non-renewed, non-renewed with direct current electrodegradation (DC-ED) and non-renewed with alternative current electro-degradation (AC-ED). Fresh 25% standard Enshi nutrient solution were changed every 3 weeks interval, with in renewed treatment, while DC- and AC-ED treatment were applied in non-renewed solutions. Significantly greater fruit yield (225.9 g plant^{-1}) was obtained from renewed nutrient solution, which was statistically similar to fruit yield in non-renewed solution with AC-ED application. Compared to renewed solution, fruit yield was decreased to about half (114.0 g plant^{-1}) in non-renewed solution while non-renewed with DC-ED produced intermediate yield between non-renewed and renewed solution or non-renewed with AC-ED. It was concluded that growth performance was greater in renewed solution followed by non-renewed with AC-ED, while it was decreased significantly in nonrenewed solution with DC-ED similar to non-renewed solution. It was also observed a similar trend in vitamin C content while brix and citric acidity was not varied. Calcium and iron concentration in the culture solution were significantly decreased in DC-ED, consequently their contents were also found lower in crowns and roots compared to other solutions used. The strawberry yield and quality can be improved through application of AC-ED in non-renewed solution [21].

3. Robotics in hydroponic systems

3.1 Robot with position-based visual feedback (RPBVF)

RPBVF was developed to act and observe the crop in NFT hydroponic systems. The focus is on the implementation of a position-based visual feedback (PBVF) algorithm in combination with a Microsoft Kinect. AmHydro 612 NFT production unit was 1.8 m × 3.65 m × 0.9 m production unit that stored 144 plants and 144 seedlings and used a closed loop water system. Above the NFT system were placed artificial lights to improve the lettuce growth. The gullies laid on an inclined table, which angle was θ, so that water flows passively to the end of the gullies. Water was collected at the end of the gullies and directed to the water reservoir, where a water pump propelled water to the top of the gullies. To manipulate the plants, the robot (**Figure 24**) was designed as a gantry with four v-groved wheels running on two inverted angle iron tracks (x-axis). On top of the gantry was a carriage that can move back and forth over the gantry (y-axis), this was perpendicular to the x-axis. On the carriage was a mechanism to move an arm up and down (z-axis), down being the negative direction. At the end of the arm was placed a two degrees of freedom gripper which opened, closed and rotated around the y-axis.

The structure is made primarily from aluminum that allows the robot to be adjusted to accommodate different sizes of NFT hydroponic systems. The x-axis was driven by a stepper motor and a chain. A timing belt transmitted the power from the carriage on the gantry to the stepper motor. The arm on the carriage was balanced by a counterweight and was driven by a stepper motor and a chain. Two linear actuators were used to open and close the gripper and the other linear actuator was used to rotate the gripper around the y-axis. All three linear actuators were driven by a 12 DCV relay board that communicates with a Phidgets interface board, which was connected to the main computer, which was running Ubuntu Server 11.04 × 64. The Kinect vision system was mounted on the carriage so that the optical axis was along the negative z-axis. All software was programmed in C++. Every hardware component communicated with its own ROS (Robotic Operating System) node. The main hardware nodes were stepper motor node, gripper node, interface board node, position node and Kinect node. The position node keeps track of the x, y and z-position of the robot and a graphic user interface was designed to provide low level control of the system. A Microsoft Kinect camera was added to the system, which produced

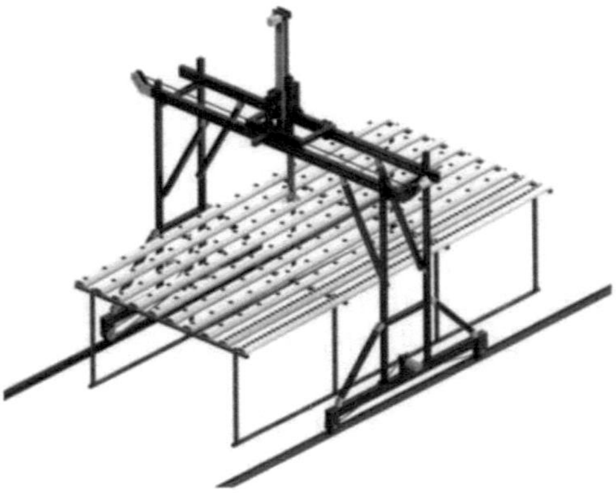

Figure 24.
Robot manipulator arm.

two kinds of images, a 640 × 480-pixel RGB color image and a 640 × 480 pixel 11-bit (0–2047) gray scale depth image that was provided by an Infra-Red (IR) sensor. The extraction of the plants required combining classical 2D image analysis techniques and IR-based depth measurement the 3D position. The computer language used was C++ using the Open Computer Vision library. The Kinect was placed on the carriage and is facing downwards (negative z-axis) to ensure the plants in its field of view are at a maximum distance of 1.5 m, because the accuracy of the Kinect decreases quadratically with distance. Up to 1.5 m the accuracy was 10 mm and the precision of the Kinect was 1 mm. The field of view was of 0.8 m × 1.15 m in x and y-direction. It was used a RPBVF algorithm was used to detect plants on the hydroponic system and placed the robot to manipulate plants (**Figure 25**).

In the algorithm, first were detected the gullies, because the plants are only located on them. All gullies are oriented along the x-axis and are straight. A probabilistic Hough Transform was used for straight line detection. By filtering the detected lines, the edges of the gullies were identified. After the identification of the edges, the lines were grouped, resulting in a segmentation of the gullies. After filtering, the coordinates of the plants in the image frame were known. Point was defined as the top left corner of the image. The depth information was extracted from the depth image by getting the value at point. The OpenNI driver transforms the IR sensor values into distances in meters by using a fitting function. To reduce the noise, multiple consecutive frames were averaged to calculate the plant coordinates. The plant coordinates form the control input for the robot. The output only depends on the current state and the control input. The open loop control algorithm was used. To be able to pick up a plant, the image frame coordinates had to be transformed to gantry coordinates. To transform the image frame coordinates to gantry coordinates a Garstka and Peters modified transformation was used. Because the Kinect is not located on the gripper, all coordinates have to be offset. These offsets are dependent on the position of the Kinect relative to the gripper. The z-coordinate has to be offset by an extra value, because the NFT table is under an angle of 2.2°. In this transformation, the point is the principal point in pixels of the depth sensor and the focal lengths in pixels were calculated. The values were quantified by calibrating the Kinect. The position bias was removed by a linear scaling of the x and y-coordinates. To evaluate the performance of the positioning and control algorithm, the x, y and z-position error between the final position of the gripper and the plant coordinate was measured. The final position of the gripper was defined as 20 mm above the center of the cup. On the top of the cup a cross-hair is drawn to mark the center. The initial position of the gripper is defined as the middle between the

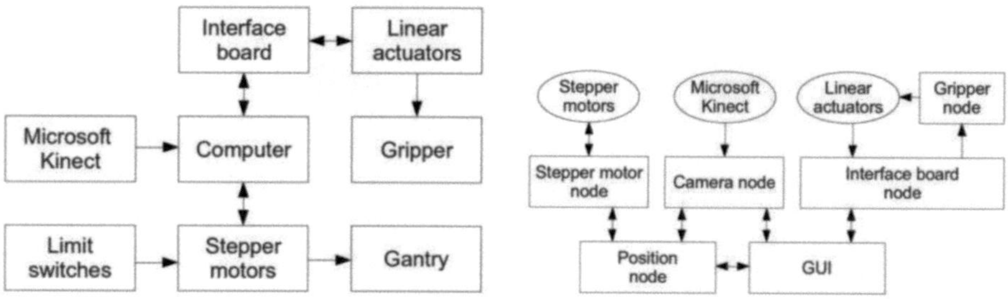

Figure 25.
Hardware layout of the robot (a) and software layout of the robot (b).

points of the gripper so the x, y and z-position error can be measured. Each image was analyzed to detect the plants. With these coordinates the robot is heading to the plant. The position error was measured with a ruler at the final stopping position and robot then returned to the same starting position. The gripper must be ±15 mm in x-direction, ±20 mm in y-direction and ±10 mm in z-direction from the center of the cup to allow the robot to pick up the plant. From the images with detected plants the gantry coordinates of the plants were calculated and inputted for the positioning algorithm so that the robot can be positioned to pick up the plants. There were 25 samples evaluated. The performance of the system is within the requirements and the plants could be manipulated on an NFT system [22].

4. Conclusions

It is expected that future developers can to detect acidity levels of pH solution, viscosity, oxygen and other variables. The future work will be collecting environmental data, which are obtained from sensors and implanting an artificial intelligence in robots and in hydroponics systems.

It is also expected that in future research make the hydroponics systems and robots able to make information panel with other operating systems that can be used as a standard system.

Acknowledgements

AILM was supported by PAICYT (CT696-19) and JMMJ was supported by PAICYT (CT571-18). We thank the Universidad Autónoma de Nuevo León, Mexican Ministry of Education, as well as Mexican Council for Science and Technology for their support.

Author details

Alejandro Isabel Luna Maldonado*, Julia Mariana Márquez Reyes*,
Héctor Flores Breceda, Humberto Rodríguez Fuentes,
Juan Antonio Vidales Contreras and Urbano Luna Maldonado
Departamento Ingeniería Agrícola y de los Alimentos, Facultad de Agronomía,
Universidad Autónoma de Nuevo León, General Escobedo, Nuevo León, Mexico

*Address all correspondence to: alejandro.lunaml@uanl.edu.mx
and julia.marquezrys@uanl.edu.mx

References

[1] Alexandratos N, Bruinsma J. World Agriculture Towards 2030/2050: The 2012 Revision. USA: University of Minnesota; 2012

[2] Resh HM. Hydroponic Food Production: A Definitive Guidebook for the Advanced Home Gardener and the Commercial Hydroponic Grower. Boca Raton, Florida, USA: CRC Press; 2016

[3] Sheikh BA. Hydroponics: Key to sustain agriculture in water stressed and urban environment. Pakistan Journal of Agriculture, Agricultural Engineering and Veterinary Sciences. 2006;**22**:53-57

[4] Kozai T, Niu G, Takagaki M, editors. Plant Factory: An Indoor Vertical Farming System for Efficient Quality Food Production. Editorial Elsevier, London, England: Academic Press; 2015

[5] Mehra M, Saxena S, Sankaranarayanan S, Tom RJ, Veeramanikandan M. IoT based hydroponics system using deep neural networks. Computers and Electronics in Agriculture. 2018;**155**:473-486

[6] Fonteno WC. Growing media: Types and physical/chemical properties. In: Reed DW, editor. Water, Media and Nutrient for Green House Crops. Batavia, IL: Ball Publishing; 1996. pp. 93-122

[7] Alipio MI, Cruz AEMD, Doria JDA, Fruto RMS. On the design of nutrient film technique hydroponics farm for smart agriculture. Journal of Engineering in Agriculture, Environment and Food. 2019;**12**:315-324

[8] Trejo-Téllez LI, Gómez-Merino FC. Nutrient solutions for hydroponic systems. In: Hydroponics—A Standard Methodology for Plant Biological Researches. London, England: IntechOpen; 2012. pp. 1-22

[9] Atmadja W, Liawatimena S, Lukas J, Nata EPL, Alexander I. Hydroponic system design with real time OS based on ARM Cortex-M microcontroller. In: IOP Conference Series: Earth and Environmental Science, Vol. 109(1). IOP Publishing; 2017. p. 012017

[10] Suhardiyanto H, Seminar KB, Chadirin Y, Setiawan BI. Development of a pH control system for nutrient solution in ebb and flow hydroponic culture based on fuzzy logic. IFAC Proceedings Volumes. 2001;**34**(11):87-90

[11] Yildirim M, Dardeniz A, Kaya S, Ali B. An automated hydroponics system used in a greenhouse. Scientific Papers, Series E, Land Reclamation, Earth Observation & Surveying Environmental Engineering. 2016;**5**:63-66

[12] Matos J, Gonçalves PJ, Torres PM. An automatic mechanical system for hydroponics fodder production. Romanian Review Precision Mechanics, Optics and Mechatronics. 2015;**47**:63

[13] Şahin İ, Calp MH, Özkan A. An expert system design and application for hydroponics greenhouse systems. Gazi University Journal of Science. 2014;**27**(2):809-822

[14] Sihombing P, Karina NA, Tarigan JT, Syarif MI. Automated hydroponics nutrition plants systems using arduino uno microcontroller based on android. Journal of Physics: Conference Series. 2018;**978**(1):012014

[15] Crisnapati PN, Wardana INK, Aryanto IKAA, Hermawan A. Hommons: Hydroponic management and monitoring system for an IOT based NFT farm using web technology. In: 5th International Conference on Cyber and IT Service Management (CITSM); IEEE. 2017. pp. 1-6

[16] Ferentinos KP, Albright LD, Selman B. Neural network based

fault detection in hydroponics. IFAC Proceedings Volumes. 2001;**34**(26):37-42

[17] Ferentinos KP, Albright LD. Predictive neural network modeling of pH and electrical conductivity in deep–trough hydroponics. Transactions of ASAE. 2002;**45**(6):2007

[18] Van LD, Lin YB, Wu TH, Lin YW, Peng SR, Kao LH, et al. PlantTalk: A smartphone-based intelligent hydroponic plant box. Sensors. 2019;**19**(8):1763

[19] Kurashina Y, Yamashita T, Kurabayashi S, Takemura K, Ando K. Growth control of leaf lettuce with exposure to underwater ultrasound and dissolved oxygen supersaturation. Ultrasonics Sonochemistry. 2019;**51**:292-297

[20] Takahashi K, Saito Y, Oikawa R, Okumura T, Takaki K, Fujio T. Development of automatically controlled corona plasma system for inactivation of pathogen in hydroponic cultivation medium of tomato. Journal of Electrostatics. 2018;**91**:61-69

[21] Talukder MR, Asaduzzaman M, Tanaka H, Asao T. Electro-degradation of culture solution improves growth, yield and quality of strawberry plants grown in closed hydroponics. Scientia Horticulturae. 2019;**243**:243-251

[22] Tanke NF, Long GA, Agrawal D, Valada A, Kantor GA. Automation of hydroponic installations using a robot with position based visual feedback. In: Proceedings of the International Conference of Agricultural Engineering CIGR-AgEng. 2012. p. C1429

Chapter 4

Application of Nanotechnology Solutions in Plants Fertilization

Daniela Predoi, Rodica V. Ghita, Simona Liliana Iconaru, Carmen Laura Cimpeanu and Stefania Mariana Raita

Abstract

Post-modern society is viewed nowadays as a technologized society, where the great solutions to human problems can be solved by the progress of technology in economics from classical industry to communications. In the last years, nanotechnology is called to play an important part in the global food production, food security and food safety in the sense that the use of nanoscale micronutrients conduced to suppressing crop disease and the relationship between nutritional status and plant diseases is investigated. Nanomaterials are capable to penetrate into cells of herbs; they can carry DNA and other chemical compounds in the cells extending the possibility in plant biotechnology to target special gene manipulation. It is important to note that the concentration, plant organ or tissue, exposure rate, elemental form, plant species, and exposure dosage (chronic/acute) affect the plant response and in particular the distinct stress response. The complex process of utilization nanoparticles in agriculture has to be monitored to a level that avoids further environmental contamination. The present and future use of nanoparticles as micronutrients is affected by different risks related to nanotoxicity of micronutrients, a problem to be solved by an appropriate and safe circuit of nanoparticles in soil, water, plants and at last in human organism.

Keywords: nanotechnology, nanoparticles, micronutrients, fertilization, nano-toxicity

1. Introduction

Post-modern society is viewed nowadays as a technologized society, where the great solutions to human problems can be solved by the progress of technology in economics from classical industry to communications. In this regard, nanotechnology was viewed from the beginning as the manipulation of matter at atomic, molecular and supramolecular scale leaved the established field of microfabrication, semiconductor physics, energy storage to extended surface science, organic chemistry and molecular biology applications. Nanotechnology presents the ability to create new materials with dimensions on the nanoscale together with a large range of applications in new domains as nanomedicine, nanoelectronics or biomaterials. From this point of view, in the last years nanotechnology is called to play an important part in the global food production, food security and food safety in the sense that the use of nanoscale micronutrients conduced to suppressing crop disease and the relationship between nutritional status and plant diseases are investigated. The large use of nanotechnology raises the problem of the toxicity of the new materials involved and their use in economics, this problem is associated with a poor legislation in the field regarding accidental release, atmospheric deposition, deliberate disposal in the environmental as pesticides, remediators, including the use of soil amendments containing nanomaterials (manuers, sludge) or water contamination for irrigation. In spite of this toxicological concern, the agriculture-nanotechnology is viewed as a solution, as a technological advancement in order to use efficiently the natural agriculture resources, i.e. water, nutrients or chemicals while farming. Besides the possible benefits of enhancing the crop yield, nanotechnology presents itself that having the ability to maximize the benefits of natural agriculture resources, through efficient products in the form of pesticides for pest and disease management and for sensors that monitors the soil quality and plant health, in the other words to solve a problem of environmental pollution. In this regard, over the last decade an important number of patents have been proposed and different products on the market that incorporated nanomaterials have been used in agricultural practice, e.g., nanopesticides, nanofertilizers or nanosensors.

As a general aspect, of world-wide society characterized by a constantly growing of population number, there exists the most important challenge that of higher agriculture yields. The aim of application of nanomaterials in agriculture is to reduce the applied amount of plan protection products, to minimize nutrient losses in fertilization and increased the yields through an optimized nutrient management. Classical active substances used nowadays can be lost during application due to different processes as runoff, evaporation, photolysis, and hydrolysis or microorganisms degradation. Different nanomaterials used as additives are characterized by large surface area and as a consequence they are appropriate to sorption process minimizing in this way the losses by reducing runoff and decreasing releases kinetics. Special designed nanoparticles can protect the active ingredients from photodegradation or can enhance uptake into leaves and other parts of the plant. Nanomaterials characteristics conduced to the substitution hazardous organic solvents present in some plant protection products and can reduce the application rates through their enhanced reactivity. As it was expressed before, despite of these positive impacts some nanomaterials have properties that classified them as potentially hazardous. The use of nanomaterials in agriculture and especially in plant protection and fertilization may pose unpredictable risks due to the fact that their application is accompanied by an intentional input of nanomaterials in the environment. In this regard, the human and environmental exposure due to nanomaterial residues in crops and soil might increase due to bioaccumulation of nanomaterials in the environment and food chain [1]. The requirements of a growing food market implied the existence of an urgent demand for products containing nanomaterials due to a process of regulation. At the beginning of twenty-first century the most popular agriculture application of nanotechnology is focused on plant protection and fertilization. It is stated [2] that higher plants have an ability to develop mechanisms to perform satisfactory under hard atmospheric and soil conditions. In order to help plants growth one of the novel methods is the use of nanomaterials that possesses physicochemical characteristics to enhance the metabolism of plants. In this view, the fertilization that used nanotechnology can amplify the plant production by delivering the micronutrients on request and control the development of plants. Nanomaterials are capable to penetrate into cells of herbs, they can carry DNA and other chemical compounds in the cells [3], extending the possibility in plant biotechnology to target special gene manipulation.

One of the most serious and important problems of agri-nanotechnology is the absence of analytical methods to quantify the concentration of nanomaterials in water, soil and air, in order to define an exposure limit. Part of difficulties is related to extraction and separation from soil matrix and interfering constituents, and the presence of very low concentrations, more over for metallic nanomaterials there exist different natural constituents as counterparts. However, the analysis techniques indicates possibility to extract these low concentrations by processes as X-ray based techniques, chemothermal oxidation, thermogravimetry or mass spectrometry, these techniques been used generally coupled on a measurement line. The evaluation risk of organic compounds used in plants protection products, namely the evaluation of persistence, bioaccumulation and toxicity is based on specific end points and parameters obtained from laboratory and field experiments. For instance, the persistence is evaluated considering the dissipation of 50% of initial concentration, bioaccumulation properties are measured of octanol-water partition coefficient and the evaluation of toxicity is based on aquatic toxicity namely the intrinsic toxicity of the compounds.

In the agri-nanotechnology the enhanced yield is related to the potential nutritional value of nanomaterials, especially for the essential micronutrients necessary for host defense. The permanent search for new solutions to global food problem conduced to the application of nanotechnology to enhance the efficiency and sustainability of agriculture practice.

2. Nanomaterials in plant growth

The increase requirement of global food production is related nowadays with the necessity of application new technology for enhancing crop yield in order to satisfy the global food security. As a modern trend in this view, the application of nanotechnology solutions can bring a response to grave problem of different deficiencies in human population as deficiencies of iron, zinc, selenium, calcium, phosphorus or vitamin A. The nanotechnology can offer solutions as micronutrients in agriculture in order to optimize the deficient presence of these substances in soil, by their use in fertilization. Besides the possible studied benefits, it is stated [1] that the nanomaterials use in agriculture may pose unforeseeable risks due to the intentional input of nanomaterials in the environment that can led to human exposure related to bioaccumulation in crops and soil and as a consequence in the food chain.

The great challenge of modern agriculture related to the use of nanotechnology is to regulate the products with the nano content in the condition where the nanomaterials pose problems to the regulatory bodies and on the other hand there is a lack of knowledge to the possible effect on the plant growth, i.e. to the genetics of plants. The possible use of nanomaterials in agriculture is a new nanotechnology solution under development now for a dozen years [4, 5] as studies regarding the use of nanoscale nutrients (metals, metal oxides, carbon) to suppress crop diseases [6]. In this view, the problem of agriculture in managing the crop disease is related to different attempts as genetic breeding, new pesticides products or new eradication protocols with the effect of the development of host plant resistance. Genetically modified plants raises different ethical problems related to the effect to the metabolism of human body, and this is a serious public concern.

A possible alternative for suppressing crop disease is the managing of plant nutrition statue and in this perspective the major limitation is that different crops have different nutrients requirements and the nutrient interacts with the level of plant disease in variable ways. As an example, the micronutrients are critical in the defense against crop disease where tissue infection induced reactions that conduced to the production of inhibitory secondary metabolites. These metabolites are generally generated by enzymes that requires activation by micronutrients cofactors, e.g. Mn, Cu and Zn as activating host defense enzymes i.e. phenylalanine or ammonia lyase. The availability of micronutrients level is related to soil characteristics, e.g. Fe, Mn, Zn are deficient in alkaline soils which limit uptake by roots and by consequence exposed roots to infections. Another way for enhancing disease defense is connected to non-essential elements, e.g. Al or Si, that offers resistance to a number of foliar and root pathogens although their presence in soil (e.g. Si) is frequently limited. As regarding Al, its application has been limited due to the fact that the over-application can cause significant crop damage and yield reduction, and insufficient presence modifies the acidity of soil [7]. The important characteristics of the nanoscale metals and metal oxides is the greatly their availability and translocation within plants. In the process of producing nanomaterials to be applied in agriculture there are used besides chemical and physical methods for synthesis the biosynthesis using plant extracts. The traditional method is the synthesis on chemical route, namely the reduction in liquid phase with common reducing agents as; citric acid, hydroxylamine, cellulose, hydrogen peroxide, sodium carbonate and sodium hydroxide. In the solutions are added stabilizing agents in order to assure uniform particle distribution and dispersion, agents such as: polyvinyl alcohol and sodium polyacrylate. The physical methods for synthesis include laser ablation, chemical vapor deposition (CVD), sonochemical reduction, supercritical fluids or gamma radiation. In the case of carbon, the fullerene synthesis is realized in arc discharge or gas combustion and carbon nanotubes are produced by CVD in the decomposition process of gaseous hydrocarbons. Different nano products that can be used as fertilizers have been patented in the last years as: active nano-grade organic fine humic (CN 1472176-A, Wu et al.); oxide nano rare earth (CN1686957-A, Wang et al.), carbon nanomaterials (US 0174032-A1, Lui et al.); nanosilver (KR 000265-A, Kim et al.); nano diatomite and zeolite (US 0115469-A1, Yu et al.), nano-selenium (US 0326153-A1, Yin et al.) or nano-silicon carrier (US 0225412-A1, Sardari et al.). The high-surface area nanoscale materials conduced to a more efficient retain of nutrients and represents a stable reservoir to plants [8], raising the potential for enhanced plants growth. The use of traditional is characterized by fertilizers with active ingredients that have low water solubility, the result being an inefficient availability to plants and furthermore a lack of control to pathogen agents. The nanofertilizers offer controlled release and synchronization of the nutrient flux over time with the uptake, minimizing the wasteful interactions with soil or air that conduced to nutrients loss. From the roots of the plants the nanomaterials as ZnO, TiO_2, CeO_2, Fe_3O_4, $Ni(OH)_2$, C_{70} fullerenes, Al, Cu, Ag, carbon nanotubes (CNT), are uptake and translocated to plant stem where partly are deposited (C_{70}, Fe_3O_4,CeO_2, $Ni(OH)_2$) or partly are foliar deposited (Al, Ag, Cu, Zn, ZnO, CeO_2, Fe_3O_4, C_{70}). The root cell of a plant has different absorption zones for different kinds of nanomaterials, for instance Fe_3O_4 has absorption areas in epidermis, cortex and cambium; $Ni(OH)_2$ in epidermis, cortex, cambium and metaxylem, Ag in epidermis and cortex or Ag^{2+} in epidermis, cortex, endodermis, and metaxylem.

Regarding nanomaterials exposure there exists a positive experience impact on crop growth and pathogen inhibitions, as related to antimicrobial activity for Ag, ZnO, Mg, Si or TiO_2. The effect on different plants of the foliar exposure to nanomaterials as ZnO conduced to increase in shoot length (15.1%), root length (4.2%) [9], increase in chlorophyll (24.4%) soluble leaf protein (38.7%) or increase in acid phosphatase (76.9%), alkaline phosphatase (61.7%) and phytase (>3x). The effect of tobacco culture cell exposure to MWCNT (multiple wall carbon nano tubes) conduced to enhanced cell growth and regulate cell division by activating water channel protein [10] and the effect of 50 μg/ml on tomato roots to MWCNT conduced to enhanced fresh and dry mass besides changes in gene expression (water channel protein) [11]. The foliar and root application of nanoparticles of Fe_2O_3 conduced to the increasing of root elongation and to the increase of photosynthetic parameters by foliar application [12]. The application of Mn in concentration 0.05–1 mg/L on Mung bean roots in a Hoagland culture solution conduced to an increase in shoot and root length, dry and fresh biomass and rootlet number [13]. The effect of spinach roots exposure to TiO_2 nanoparticles present in soil conduced to an enhanced growth rate and chlorophyll as well as an enhanced rubisco activity and photosynthetic rate [14]. The silver nanoparticles exhibit an intense inhibitory activity to microorganisms, in this regard Ag NPs damaged and penetrate the cell membrane subsequently reducing the infection [15]. Another nanomaterial with intense antimicrobial activity is ZnO NPs that is effective to pathogen control growth, also characterized by a lower toxicity in comparison to Ag and with benefits on soil fertility. The application of ZnO NPs conduced to systemic disruption of cellular function of pathogens as *Botrytis cinerea* or *Penicillium expansum* resulting in hyphal malformation and fungal depth [16]. Another promising amendment is TiO_2 NPs due to their combined photo-catalytic and antimicrobial activity, e.g. application of TiO_2 NPs reduced *P. cubensis* infection of cucumber by 91% and increase photosynthetic activity by 30% [17]. One of the most popular cultures is that of wheat, in this case the foliar application of Ti NPs at 20 g/l it increased stem elongation, biomass, flowering, ear mass and seed number [18]. Nanomaterials fertilizing activity is influenced by the chemical and physical characteristics of the environment soil, air and water. The initial properties of NM can suffer different transformations due to the interactions with both biotic and abiotic soil components and these modifications can influence the stability of NPs, transport and aggregation and availability to plants. Necessary micronutrients as Cu, Fe, Mn or Zn become less availability from the soil when pH is approaching to the limit of 7.0 and subsequently there exist a lower uptake to plants roots and compromising the nutritional status [19]. It is reported [20] a pH-dependence of humic acid adsorption onto nanoparticles of TiO_2, Al_2O_3 and ZnO where the electrostatic interactions and ligand exchange with SiO_2 is responsible for selective adsorption onto oxide surface. Due to different limitations related to soil characteristics and their content in macro and micro nutrients, it became important the foliar applications of fertilizers, in particular the nanomaterial nutrients. The entrance gate for micronutrients is the leaves through stomata and cuticle structures as the studies in literature have presented [21]. In case of watermelon [22] the pathway leaf-to-root translocation of nanomaterials after a foliar application presented an important content of Ti, Mg or Zn nanoparticles that exist in root tissue a fact that shows the effective action of foliar application of nanomaterials. The possible route of biosynthesis of nanoparticles, e.g. ZnO by extracellular secretions of *Aspergillus fumigatus* TFR-8 and their applications as a foliar spray conduced in case of bean plants [23] to an increase of physiological parameters i.e. biomass, shoot/root length, root area or chlorophyll content and on the other hand the residual protein from fungal extract increased nanoparticle stability. Regarding TiO_2 nanomaterials is stated [24] that the application on different crops, e.g. wheat or soybean has increased the yield and reduced the pathogenic diseases, these effects being based on surface properties of TiO_2 nanoparticles as their photo-catalytic characteristics. Another nanomaterial applied to cucumber leaves is CeO_2 [25] in nano-powder form, showed a high leaf-to-root translocation suggesting a phloem-based transport throughout the plant. Foliar applied nanoscale amendments dedicated to pathogen control are related to antifungal activity of CuO_2 nanomaterial or Ag nanoparticles. Tested on tomato infected with *Phytothphora infestans* the effect of CuO_2 nanomaterial [26] showed an increased protection (73.5%) as compared to bulk amendment (57.8%) promoting the use of nanoscale amendments both to suppress disease and enhanced the nutrient action of nanomaterials. The antifungal activity of Ag nanoparticles is based on the accumulation in the fungal hyphae that disrupted cellular function, an intense process related to a higher ion release on the increased nanoparticle surface area [27]. It is worth to mention that the problem of *in planta* translocation, i.e. the way that the foliar application of nanoscale nutrients affects root pathogens, is still under research in the sense that pathogens can be released after shoot-root transfer or the induced host resistance. A non-classical nutrient besides metal and metal-oxides there is based on carbon nanomaterials in the forms of $C_{60/70}$ fullerenes, carbon nanoparticles, or single/multiple wall carbon nanotubes (SWCNT/MWCNT). An extended study [11] in literature upon the action of MWCNT, SWCNT, graphene and bulk activated carbon onto tomato plants grown in artificial medium revealed an enhancement of biomass by stimulating the growth. The molecular analysis upon the action of MWCNT has shown a stimulation of cell division and plant growth due to the activation of water channels (aquaporins) and regulatory genes for cell division and extension. Carbon nanomaterials exposure can alter the different co-existing organic contaminants in various kinds of soils. In this regards, carbon nanomaterials presents toxicity to soil microorganisms, with accent to SWCNT including fungal community. Carbon nanomaterials have potential to enhance plant growth, nutrient uptake, seed germination or fruit yield the most promising one being MWCNT with positive effects on different crop species. The large inters in the use of nanomaterials is based on the increase global production of nanomaterials and their possible application in agriculture with hazards and risks to be investigated. An exposed [28, 29] "realistic exposure scenario" for TiO_2, Ag and carbon nanotubes proposed the doses of 0.4, 0.02 and 0.01 μg/kg/year although the relationship between these values and the actual concentration in the environment is not known.

It is worth to mention, that in general the discussion to nanomaterials in agriculture refers also to a most prominent fraction of nanomaterials that are non-solids comprising nanoscale structures that can encapsulate an active ingredient in plan protection product. Generally active substances have poor solubility in water and at room temperature are brought to solution with organic (co)solvents. In order to avoid the use of organic (co)solvents one solution is stated [30] the use of oil/water emulsions. Generally the physical appearance of non-solid nanomaterials are lipid base in liposomes, micelles or cochleates, in polymer based in micelles, nanosphere, nanocapsules and polymersomes or in emulsions base as liquid crystals and micro-emulsions. The nanomaterials in non-solid forms enhance the solubility and the coverage of the hydrophobic leaf surface together with the penetration of the active substances through the cuticula.

As presented, the characteristics of solid and non-solid nanomaterials have been investigated in the last decade in order to understand the effect of nanonutrients in culture fertilization as well as in plant protection with promising results together with various studies regarding the toxicity of nanoparticles in the environment.

3. Nanoparticles and their action on plants

The nanotechnology application in agriculture is for our world one of the most important domains of study due to the possibility of increasing for different culture production and assisted plant protection against pests and diseases together with the monitoring of pathogenic agents. In this regard, the application of nanotechnology in the control of crop yield and crop protection is relatively recent compared to organic or chemical nutrients and different drug delivery or pharmaceuticals [31, 32].

One of the most important nutrients for humans and plants is iron (Fe) where iron deficiency is common in nowadays human diet affecting over 2 billion people in the world [33] Iron is essential for plant development and plays an important part in photosynthetic process being implied in redox reaction as well as generating reactive oxygen species [34]. Due to its properties iron containing nanoparticles have been used as nano-fertilizer for nutrition of plants. As an example there was observed [35] a positive effect of nano-FeO and nano-ZnCuFe oxide on the growth of mung (*Vigna radiata)* seedling, as well a positive influence on leaf and pod dry weight on soybean yield and quality [36]. The problem of high rate of accumulation of iron oxide nanoparticles in plants that conduced to precipitation in gravitational field can be solve by surface-coating materials [37], a promising way to improve the agronomic traits. Surface iron coating materials as nano-Fe_2O_3 coordinated with humic acid improved the mobility of iron in peanuts [38] or water-based ferrofluids stabilized with citric acid on the growth of maize [39]. In this spirit, in literature [40] there are data regarding the action of nono-iron fertilizer capped with ethylenediaminetetra acetic acid (EDTA) upon sunflower(*Helianthus annuus)* by foliar and soil application. An important parameter in studying the functional biology is plant biomass, in this regard the effect of applied nano-fertilizer by foliar treatment is the improving of aerial organ dry biomass while the effect on soil application of nano-Fe-EDTA is not conclusive. Regarding aerial organ fresh biomass both foliar and soil addition of nano-Fe-EDTA had the effect of increasing aerial organ fresh weight of the plants as compared to classic fertilizer. In this method of applying nano-Fe-EDTA there exists an increasing of leaves number in percentage by 21.42% as compared to control plants [40]. As regards, the plant height the application of nano-Fe-EDTA is effective also in foliar and soil treatment and taking into account these experiments the general growth of sunflower is influenced by iron oxide nanoparticles. The nano-fertilizer that treated the plan influenced also the physiological parameters in the sense that all treated plants had a higher chlorophyll content, the most important pigment, level than the test plants i.e. from the total level of 2.69 mg/g and 2.34 mg/g belongs to chlorophyll to the treated plants with nano-Fe-EDTA through soil absorption, indicated the translocation of coated nanoparticles from roots to the aerial parts. This treatment method implied penetration points on the leave surface i.e. stomata and subestomatic chambers which means hydrophobic penetration of nanoparticles through these pores. Another problem for foliar application of nanoparticles is the possibility for their accumulation in cells from epidermis of petiole near the application point limiting in this way their possible contribution to plant growth or photosynthesis reaction. However, the effect of an organic shell around nano-Fe made these nanoparticles more compatible for entering and translocation in the plant [40]. Another direction to suppress iron deficiency common in human diet is to use iron fertilizers based on humic substances extracted from lignites, such as leordine, it is stated [41] that this kind of fertilizer is more ecofriendly than synthetic iron chelates as discussed above, but they are less efficient in suppressing iron chlorosis. Low concentrations of superparamagnetic Fe-nanoparticles increased significantly the chlorophyll contents in sub-apical leaves of soybeans under hydroponic conditions. The plants fertilized with the leonardite humates accumulated slightly higher fresh weight than those fertilized with the iron chelate, the humic substances generally increase the shoot and root growth by 15–25% and the accumulation of total iron in pods for soybean plants reaches 50 mg/kg under conditions of sufficient nourishment [41]. The applied nanoparticles of Fe^{57} were capable to supply the Fe^{57} deficiency in plant and it was transported from root to shoot and reaches the pods, this iron humate was prepared taking into account its maximum complexing capacity in order to avoid the iron flocculation in calcareous conditions [41]. As a remark, in the context of sustainable agriculture, the Fe-nanoparticles can be considered as a part of novel technology in line with the politics of precision and sustainable agriculture.

One of the elements that results from the rapid industrialization is cadmium-Cd and as a consequence there exists an irreversible exposure in the environment, especially in the soil. The Cd absorption in the plants from soil or air through aerial deposition and its transfer into different parts of the plants can cause several abnormalities in plants as reduced growth and yield [42]. The major entry gate of Cd in plants is the roots while the toxic element entry in the human body is the consumption of contaminated food. The excess of Cd in plants affected the plant growth by reducing the production of reactive species, electrolyte leakage, hydrogen peroxide and malondialdehyde concentrations in plants [42]. As a solution to reduce the Cd content in soil is the application of biochar, as a carbon rich pyrolyzed organic biomass that is effective in reducing bioavailability of metals in soil [43]. These properties are based on biochar high pH, cation exchange capacity, nutrient retention capacity including water retention capacity and lower bulk density [44]. The use of nanotechnology in agriculture can rise different problems as the role of foliar application of ZnO nanoparticles combined with soil applied biochar in Cd accumulation by plants [45], in this regard it was stated that compared to other cereals maize (*Z. mays*) plant has a higher ability to take up Cd and its translocation to the aerial parts that conduced to Cd accumulation in grains. The effect of applied ZnO nanoparticles alone or combined with biochar enhanced the chlorophyll concentrations and gas exchange parameters in leaves of maize [45]. On the other hand, the effect upon on malondialdehyde, hydrogen peroxide, electrolyte leakage and antioxidant enzyme activities in maize leaf and roots are in the sense that the nanoparticles reduced these contents applied alone or with biochar. Experimental application of ZnO-nanoparticles improved the activities of antioxidant enzymes in leaves and roots [45]. As it was presented [45] the application of ZnO nanoparticles and biochar reduced the Cd concentration in maize shoots and roots. Generally the revealing that ZnO nanoparticles with effect in maize biomass and growth is expressed by accelerated exogenous application of nanoparticles further enhanced with biochar application in combination to nanoparticles. It was observed [45] that the lower biomass in control plants is associated with higher Cd concentration in maize which reduced the chlorophyll concentration in leaves, or due to the increased levels of malondialdehyde, hydrogen peroxide and electrolyte leakage in the belowground and aboveground tissues. Lower concentration of ZnO nanoparticles have positive impacts on plants [46] in the sense that the height of the plants increased and the biochar application decreased the soluble Cd in soil, meanwhile nanoparticles increased the Zn level in the plants [45]. Exogenous application of ZnO nanoparticles improved the chlorophyll concentration and as a consequence improved photosynthesis and with applied amendments might reduce the oxidative stress in maize plants. It was suggested [47] that ZnO nanoparticles can be considered as slow-release for Zn fertilizers which is advantageous to avoid sudden absorption of Zn by plants as the higher concentrations of Zn absorbed by the plants also conduced to toxicity in plants. It was stated [48] that a proper amount of Zn in the soil or plants may interfere with Cd and could reduce the Cd accumulation by plants due to the antagonistic effects of these metals on each other and furthermore biochar application in combination with ZnO nanoparticles further deceased Cd concentration in maize plants. As a solution to Zn malnutrition, the strategy of using ZnO nanoparticles combined with biochar in cereals growth, in particular in maize, conduced to the enhancement in plant biomass with decreased Cd concentration in cereals. A proper concentration of Zn is benefic to plant organism because Zn is necessary for the activity of enzymes such as dehydrogenases, aldolases, isomerases, transphosphorylases and RNA and DNA polymerases, as well as in synthesis of tryptophan, cell division, maintenance of membrane structure and photosynthesis and acts as a regulatory cofactor in protein synthesis [49].

For example in coffee plants although Zn is required for optimal metabolism, yet deficiency is prevalent partly due to the inefficient absorption of this micronutrient combined with a deficiency in translocation [50]. Zn fertilization improves the production and quality of coffee beans by positive impact on polyphenol oxidase activity, color index, sucrose content, caffeine and trigonelline content and chlorogenic acid [49]. One of the most efficient ways of suppressing Zn deficiency is foliar fertilization a method that avoids toxicity symptoms and reduce fertilizer-related pollution and in this perspective the foliar fertilization using micronutrients as Zn nanoparticles is the advanced solution proposed by nanotechnology. In spite of positive physiological impacts of nanoparticle fertilizers on crop growth, the properties of nanoparticles can induce oxidative stress and toxicity in plants and other organisms in ecosystem [51]. It was observed [52] the effect of ZnO nanoparticles application on coffee plant on its positive part, i.e. the fresh weight of roots and leaves are increased in percentages of 37% (root) and 95% (leaves) and no effect on the stem as compared to control. The net photosynthesis rate did not vary over time for ZnO nanoparticles treated plants and as regards zinc assimilation ZnO nanoparticles treated leaves contained a higher content of Zn compared to classical nutrient $ZnSO_4$ treated plants [52]. As regards, the zinc assimilation in stems and roots there does not exist significant differences between the treated plants with ZnO nanoparticles and zinc sulfate at the same content of Zn in the soil. The treatment of coffee plants with ZnO nanoparticles positively affected plant biomass, with a major effect on fresh and dry weight. Part of so-called photosynthetic machinery was improved when coffee plants were exposed to ZnO nanoparticles, there exists a positive interaction between ZnO nanoparticles and net carbon assimilation rate and stomatal conductance, confirming the role of Zn as a cofactor of carbonic anhydrase that increases the content of CO_2 in the chloroplast and thus also increases the carboxylation capability of the Rubisco enzyme [53]. It is important to state that although Zn is an essential element, in an impropriate quantity it can reduce plant health and performance at phytotoxic concentrations. Symptoms of Zn toxicity are reduced growth and plant biomass, inhibition of cell elongation and division, wilting, curling and rolling of young leaves chlorotic and necrotic leaf tips and root growth inhibition [54].

The effect of nano-boron (B) fertilizer on the mineral nutrition and fruit yield was put into evidence by on pomegranate trees culture [55]. It was observed that a foliar spray of nano-B (concentration 6.5 mg BL^{-1}) in combination with nano-Zn increased significantly the fruit yield up to 34% depending on treatment, with an accent to nano-B fertilizer. Also, the foliar application of these nano-fertilizers increased the number of fruits per tree, without an effect upon fruit cracking. As regards the fruit size physical parameters as fruit diameter, fruit calvyx diameter or average weight, they are not affected significantly for the treated trees, but the pH pomegranate juice increased to 0.62 pH units under fertilization. Concentrated nano-B foliar application caused small (1%) but statistically constant changes in the amount of total phenolic compounds in pomegranate juice whereas the antioxidant activity is not affected. The total amount of sugar in pomegranate fruit juice increased up to 4.6% at discussed nano-B concentration with no significantly increase in total anthocyanins under treatment. The action of nano-B fertilizers is also efficient in fruit crops as almond, apple, pear, persimmon or peach.

Calcium (Ca) is an important macro-element that plays an important role in plants including structural functions of cell walls, stabilization of cell membrane, maintenance of cell turgor pressure and counter-ion for inorganic and organic anions in vacuoles, as well as cytoplasmic messenger. Calcium cannot be transferred through from the older tissue to other parts of plant on the basic phloem pathway and Ca xylem translocation depends on unidirectional transpiration stream [56]. Studies of foliar application of nano-Ca on pomegranate trees shows no significant effect on fruit yield and to the number of fruits per trees [57]. Nano-Ca fertilization increased the Ca leaf concentration, whereas the foliar treatment decreased significantly pomegranate fruit cracking. The total phenolic compounds in pomegranate fruit juice is decreased in nano-Ca fertilization but with no significant effect on antioxidant activity and total anthocyanin content. Foliar application of Ca reduces the fruit cracking due to its role on cell wall, in enhancing the mechanical properties of plant tissue. However, the foliar fertilization of Ca had no significant effect on yields for kiwifruit, strawberry, grape and cherry.

Manganese (Mn) is a micronutrient required by most of the plants die to its implication in biochemical reactions, as those required by dehydrogenases, decarboxylases, kinases, oxidases, peroxidases enzymes and to their role in fighting oxygen reactive species in plants. Plants required 20–40 mg Mn/kg of dry weight for its various functions e.g. in tricarboxylic acid cycle, oxidative and non-oxidative decarboxylation reactions and for different synthesis as carotenoids, sterols or gibberellic acid. The most important process of photosynthesis, implied the final conversion of absorbed light to energy via enzymatic reactions. Among them, a studied Mn-containing enzyme, is found in PSII oxygen evolving complex, a multi-step enzymatic pathway where Mn is required, as a cofactor in both lower and higher plants for the Hill reaction-the water splitting and oxygen evolving system [58]. Mn plays an important role in the synthesis of fatty acids and carotenoids, as well as in cell division and elongation. A normal function of the plant as biological system is affected by abiotic stress defined any adverse force or condition that affects its normal functioning. Abiotic stresses as drought, flood, salinity or harsh temperature conduced to an excessive amounts of reactive oxygen species that potentially injure proteins, membrane lipids, carbohydrates and DNA. The application of Mn increased the leaf area, photosynthesis rate and stomatal conductance in drought stress conditions, reducing the production of reactive oxygen species in plants. These reactive species also accumulated in inside the plants under thermal stress as harsh temperatures, that causes damage to cellular compounds and metabolic processes. It was suggested [59] that salinity inhibits the uptake of Mn inside plants inducing deficiency, in this case the foliar application increased stem diameter, fresh and dry biomass, number of seeds and different biochemical parameters as total protein or Hill reaction activity. The application of Mn nanoparticles on wheat applied by foliar exposure or soil amendment showed [60] no inhibition of vegetative or reproductive development, further more Mn nanoparticles significantly reduced Mn accumulation in shoots but increased the translocation efficiency in grains compared to classical Mn fertilizers due to a greater reactivity and non-toxicity due to a slower and a continuously availability of soluble Mn from Mn nanoparticles as compared to ionic Mn salt. The Mn nanoparticles greater photophosphorylation and oxygen evolution compared to bulk Mn suggested its novel potential nano modulator of the photochemical pathway in agriculture [61]. Furthermore Mn nanoparticles are viewed as a stimulant of plant growth and of metabolic processes as alkaloids production. In infested soils, foliar exposure to Mn oxide nanoparticles reduces disease up to 28% as compared to controls [62], the plants can have cross-tolerance between abiotic and biotic stresses, and in this regard stress tolerance can help plants to form faster and more resistant manner to additional environmental changes. The propose mechanism for the distribution of Mn nanoparticles through the plant incudes transportation from the root through the vascular system, a process considered as active-transporting as long as includes signaling, recycling and plasma membrane regulation. Generally the excess amounts of Mn in plants is not benefic to their health by interfering with the uptake, transport and utilization of other minerals as Ca, Fe or Mg as being competitive for the same ion transport.

Copper nanoparticles at low concentrations (<0.25 mg L^{-1}) stimulated plant photosynthesis in a percentage of 35% on waterweed (*Elodea. densa planch*) compared with control plants [63]. It was stated [64] that soil amendments with metallic Cu nanoparticles up to 600 mg kg^{-1} increased significantly 15-day lettuce seedling growth up to 91% without toxic effects. Higher concentrations up to 1000 mg L^{-1} of metallic Cu nanoparticles conduced to toxic effects on seedling growth of mung bean and wheat [65] and can reduce the biomass of zucchini by 90% as compared to control. The optimal concentration for aqueous copper for proper plant growth is in the range of only 0.02 mg L^{-1} due to the effect of phytoxicity at higher levels [66].

Magnesium (Mg) is essential for plant growth as it plays an important role in the photosynthesis process as central component of chlorophyll and also acts as a phosphorus carrier as an important element for phosphate metabolism. Mg is necessary in cell division and protein formation in activation of several enzyme systems and is essential for plant respiration. The effect of foliar application of magnesium and iron nanoparticles solutions upon black-eyed pea (*Vigna unguiculata*) combined in a concentration of 0.5 g L^{-1} enhanced the 1000-seed weight by 7% in comparison with regular application of Fe and Mg [67]. The Mg nanoparticles application improved the uptake of Mg in plant stems and leaves compared to the use of regular Mg salts, a fact related to the higher mobility of Mg nanoparticles. Magnesium deficiency conduced to a slow growth of plant and to leaves problems due to the development of internal chlorosis and that is why Mg important in crop yield.

Silver (Ag) is known as an element with antiseptic characteristics and it is important to understand its effect upon soil microbial community. Microorganisms are key regulators of biogeochemical recycling of nutrients in the environment and assist in maintaining the overall health and function of ecosystems [68]. The soil is regarded as a complex system and its physicochemical characteristics as pH, texture and organic matter content can influence the nanoparticles properties introduced in it, a fact that can conduce to an increased or a decreased bioavailability and toxicity of nanoparticles. The effect of Ag nanoparticles on the microbial diversity and enzyme activity of soil is regarded as a significant decrease in microbial mass as a function of increasing Ag nanoparticles concentration. [69]. Ag nanoparticles had impact on vascular plants, presenting positive or negative effects on seed germination, root growth and plant biomass [70]. In Ag nanoparticles application on wheat plants, there was not observed a significant effect with the exception of root fresh weight and root length that presents a negative response at 75 ppm treatment while in cowpea and *Brassica* there was observed a positive response to Ag nanoparticles. In cowpea plants a 50 ppm concentration of Ag nanoparticles conduced to growth promotion and increased root nodulation suggesting that Ag nanoparticles treatment improved the growth by modulating the antioxidant action of nanoparticles. Increased nodulation is supposed to be related to nitrogen-fixing bacteria as long as root exudation pattern is dependent to Ag nanoparticles concentration. In soil, total bacteria count improved in 50 ppm treatment and nitrogen fixer bacteria are sensitive toward 75 ppm treatment. In cowpea, the total bacteria count declined with increasing Ag nanoparticles concentration, with an increase of diversity index of total bacteria population was observed in 50 ppm treatment whereas the diversity of nitrogen fixers decreased in the 75 ppm treatment. The impact of Ag nanoparticles on soil bacteria diversity is dependent on Ag nanoparticles concentration and on the other hand on the plant species grown in that soil, a specificity related to the different root exudation patterns of different plant species. The antimicrobial properties of Ag nanoparticles may be altered when released in soil due to the complex system of biotic and abiotic processes, e.g. pore-water harbors a range of electrolytes that increase the aggregation of Ag nanoparticles in soil, thus reducing its size-dependent toxicity. The plants cultured in artificial soils compared to agar conditions in addition to higher concentration of Ag^+ ions the plants exhibited some characteristics: (a) the apparent toxicity observed in the soil was attributable to the particle toxicity; (b) lower rates of nanoparticles dissolution can be attributed to the reduction in the surface and greater soil aggregation; (c) the agar and soil have different mechanisms for sorption of the dissolved Ag^+ ion and the Ag nanoparticles. The application of Ag nanoparticles in real soil improved the bactericidal and fungicidal effectiveness of silver against most important plant pathogenic fungi.

Titanium oxide (TiO_2) mineral is naturally occurred in four natural polymorphs: akaogiite (monoclinic), brookite (orthorhombic), anatase (tetragonal) and rutile (tetragonal). TiO_2 nanoparticles are explored due their use as an antimicrobial agent and photocatalyst in order to remove organic compounds from contaminated air, soil, and water. Ti is not an essential element for plants, therefore TiO_2 nanoparticles are not viewed as plant nutrients but plays a potential role in plant protection and at lower doses its effective to its properties as a photocatalysts or an UV protector. It was stated [71] that exposure of naturally aged spinach seeds to TiO_2 nanoparticles (rutile) at concentration of 250–4000 mg L^{-1} significantly increased the germination rate, the germination index, the dry weight of seedlings and vigor index of seeds. It was observed that under hydroponic conditions on agar, TiO_2 nanoparticles generally cause positive or non-consequential effects on plant growth for different food crops. For example, in hydroponic conditions it was observed a significant increase in the root and shoot length of *Brassica juncea* seedling treated with TiO_2 nanoparticles at concentrations: (0, 200, 500, 1000 and 1500 mg L^{-1}). An important factor is the lack of toxicity of TiO_2 nanoparticles due to their rapid agglomeration and consequently formation of larger hydrodynamic particles that are not available to plant and have no effect on plants, this is a property of rutile form that presented the characteristic of lipophilicity. It was observed [72] that TiO_2 in anatase form are toxic at high concentrations and due to their antimicrobial properties a significant growth of the roots was observed.

Selenium (Se) is an essential trace element for humans and animals and is beneficial for plants at low concentrations particularly under stress conditions acting like an antioxidant. Cereals are a good source of Se as it is present in the form of Se-methionine [73]. The total concentration of Se in most of soils is 0.1–2 mg kg^{-1}, with factors that affects Se solubility as: soil pH, redox potential, calcium carbonate level, cation exchange capacity, organic carbon, iron and aluminum level as well as the plants capability to produce root exudates. Selenium is most available in alkaline soils in the form of selenite, in acidic poorly aerated Se occurs in insoluble selenide forms, the lower limit for Se concentration in soil is 0.5 mg kg^{-1}. Selenium fertilization rate and its chemical form directly influenced Se grain concentration affecting the yield, its application form being foliar or liquid on the soil surface. Low concentrations of Se had a positive effect on growth of ryegrass, lettuce and potato due to its antioxidant action. It was stated [74] that 10 g Se selenite per ha can increase wheat Se concentration from a base level of 30–100-200 μg kg^{-1} to recommended level of 300 μg kg^{-1} as a minimum target. Agronomic use efficiency is higher for foliar application than soil application. Agronomic bio-fortification is an inexpensive method for the increase of Se intake by humans but the limited Se resources indicated that fertilization strategies had to increase agronomic use efficiency in environmental conditions by improving agro-technical measures.

Silicon dioxide (SiO_2) is a form of silicon oxide that had abundance in environmental mostly in the soil. Lower amounts of nano-SiO_2 increased germination of seeds in tomato [75], or of *Lycopersicum esculentum* seeds germination in concentration of 8 g L^{-1} nano-SiO_2 for a percentage of 22.16%. The same concentration increased the fresh weight of seedlings by 116.58% and seedlings dry weight by 117.46% compared to control, with an important action upon root and shoot growth. Nano-SiO_2 amplified various factors of the growth and conditions of seedlings, i.e. height, diameter of root collar, main length of roots, seedlings lateral root number as well as induction of chlorophyll synthesis. Under abiotic stress nano-SiO_2 increases seeds germination in tomato [76] and in saline conditions nano-SiO_2 maximized the fresh and dry weight of leaves confirmed that the increment in the proline accumulation, unattached amino-acids, nutrient quantity and antioxidant activity of enzymes increased plant's level of endurance to environmental stress [77]. The chlorophyll content developed on nano-SiO_2 treated plants grown in salt-stressed conditions, the application of 1 mM silicon dioxide could alleviate the side effect of salt stress on percentage of germination, length, of root and shoot, weight of seedling, mean germination rate see vigor index and cotyledon reserve mobilization of *Lens culinaris* [78]. Nano silicon dioxide developed the growth of the plant, net rate photosynthesis, level of transpiration, conductance of stomata, rate electron transport and photochemical quell [79].

Regarding the nanotoxicity of nanoparticles action, the involved mechanism is not entirely understand, this mechanism is assumed in the sense by the changes related to the chemical structure, particles size and active surface area of nanoparticles. It is stated [80] that the toxicity action is focused on two directions: i.e. (a) a chemical toxicity based on the chemical composition as the released of toxic ions and (b) stress or stimuli caused by the surface, size or shape of particles. As viewed from chemical physics processes, the solubility of oxide nanoparticles affects the cell culture response and nanoparticle mediated toxicity is partly explained by the release of dissolved components of them [81]. In comparison to metal toxicity in plants and animals, the nanoparticles pathway is different, a problem solved by the introduction of different parameters in experimental tests to evaluate the nanoparticles dynamics. In a plant culture exposed to nanoparticles, the gain and losses related to the development, growth and productivity are not exclusively part of nanoparticles effect but they can be viewed as a participation of the primary ions to biological processes active in plants. Regarding the presence of nanoparticles in soil, as culture area for plants, had to be considered the interaction with the microorganisms in soil because they can positively interact with plants e.g. arbuscular mycorrhizal fungi [80]. The nanoparticles interaction with plants e.g. accumulation in plant biomass affects their fate and transport in the environment. There exists a first report [82] in 2007 regarding the negative effects of nanoparticles upon several plants as corn, cucumber, soybean, cabbage and carrot at relatively low dosage. At microscopic scale, the analysis of the chromosome morphology showed a relation between increased number of aberrations and the increased concentration of nanoparticles e.g. the appearance of stick chromosomes [83]. The phyto-toxicity was observed at molecular and nuclear cell level because the occurrence of stick chromosomes might be related to the degradation or de-polymerization of chromosomal DNA [80]. The extended development of modern agriculture brings besides some true benefits regarding crop productivity increasing problems related to environmental contamination with toxic elements e.g. metals or pesticides compounds and from this point of view nanoparticles used can aggravate the situation. The process of heavy metals accumulation by plants is related for the large majority of plants to roots accumulation and only a small part is translocated to the aboveground of the plants [84]. There are evidence [85] to a translocation process from the roots to the fruits, without the existence of changes due to biochemical processes. It is important to note that the concentration, plant organ or tissue, exposure rate, elemental form, plant species, exposure dosage (chronic/acute) affects the plant response and in particular the distinct stress response. This is the reason why the complex process of utilization nanoparticles in agriculture has to be monitored to a level that avoids further environmental contamination i.e. soil, water and air.

4. Hydroxyapatite and essential oils

One of the essential problems that limits crop production is the low availability of phosphorous (P) in many agricultural regions. In order to increase the phosphorous content in the soil it is necessary to apply P fertilizers, because phosphorous is an essential element for plant growth. Besides classical fertilizers, nanotechnology can offer a solution to supply the micronutrients deficiencies for plants development. In this regard, it was suggested [86] the use of hydroxyapatite nanoparticles $Ca_{10}(PO_4)_6(OH)_2$, nano-HAp, as potential fertilizers. Tests on the potential use of nano-HAp (15 nm) on soybean (*Glycine max*) [87] conduced to the conclusion that plants which received the nano-fertilizer produced greater biomass yield and seed compared with the plants fertilized with conventional water-soluble phosphorous source. It was stated [88] that nano-HAp might enhance P fertilizer efficiency in acidic and strongly P absorbing soils through the better mobility of nano-HAp in the soil potentially reaching plant roots. The experiments showed that the application of conventional triple superphosphate and HAp treatment conduced to an significantly increased the plant dry matter yield compared to control depending on soil characteristics e.g. Greenwood soils. The application of conventional fertilizer conduced to significant increase in shoot uptake of phosphorous in all types of soils in the range (0.1–1.4 mg P plant^{-1}), in peculiar the application of nano-HAp increased the plant uptake to a maximum value of 0.4 mg P plant^{-1}, indicating a contribution uptake in percentage (40–61)%. These results suggesting that nanoparticles could possibly have a benefit over water-soluble P conventional fertilizers in the very strongly absorbing soils since the opportunities for P fixation in the soil is minimized.

Among other important plants for the use in medicine are those conducted to the extraction of essential oils as basil and lavender. Essential oils are complex biostructures and contain a lot of chemical compounds from different classes as: terpenoids, ketones, aldehydes and esters with a composition depending on the plant's origin and quality, harvest time, climate, soil and extraction method. About 90% of the bioactive components of essential oils are monoterpenes and in the oxygenated form the bioactivity is enhanced. The medical use of essential oils is based on their properties such as antibacterial, antifungal and their characteristic to prevent the growth of different pathogens. The antibacterial activity of HAp samples and HAp samples coated with essential oil of basil and lavender was tested on methicillin_resistant *Staphylococcus aureus* (MRSA), *Staphylococcus aureus* 0364 and Gram-negative bacteria *Escherichia coli* ATCC 25922 as presented in [89]. As observed in **Figure 1(a–c)** there exists a slow decrease in MRSA growth for HAp-B and HAp-L samples at different concentrations—**Figure 1(a)**.

The evolution in the cell growth of *S. aureus* is observed in **Figure 1(b)**, an evolution that indicates a decrease in the cell growth starting from the low level concentration of 0.01 mg mL^{-1} in the presence of the HAp-B sample. The HAp-L inhibited *E. coli* (**Figure 1c**) starting with the concentration of 0.02 mg mL^{-1}. Taking into account that MIC is the lowest concentration of the chemical that prevents visible growth of bacteria, the lowest concentration of HAp-L nanopowders at which the visible inhibition of MRSA bacterial growth was observed was 0.039 mg mL^{-1}, and for HAp-B was 0.625 mg mL^{-1}. In the same conditions for *S. aureus* we have for HAp-B MIC value of 0.313 mg mL^{-1} and for *E. coli* HAp-L

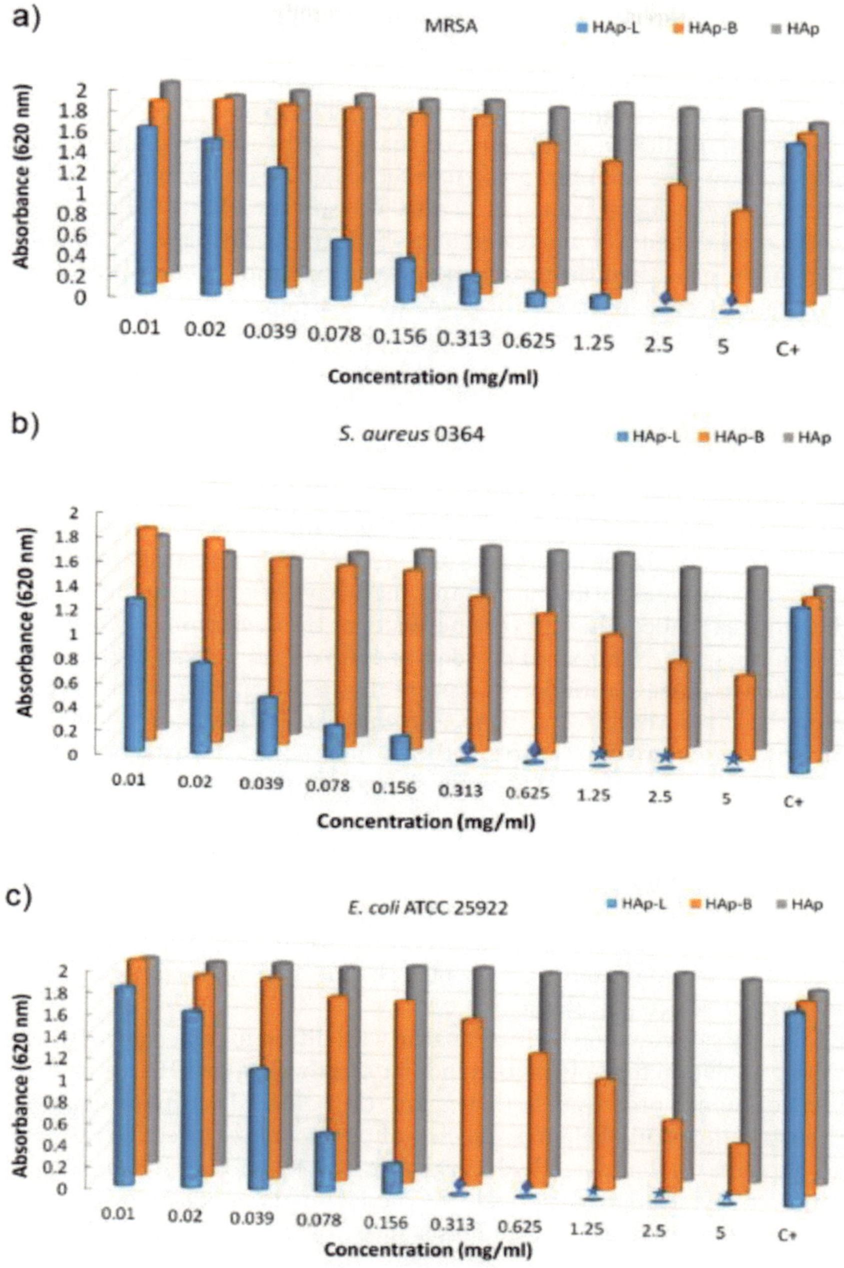

Figure 1.
Graphic representation of absorbance values of the microbial culture obtained in the presence of the plant EO-coated HAp on different bacterial strains MRSA (a), S. aureus *(b) and* E. coli *(c) quantified by A620 nm values.*

MIC value was 0.039 mg mL^{-1}. The antimicrobial properties of materials based on hydroxyapatite coated with essential oils can be explained for HAp-B samples on the major component of hydroxyapatite as the basil essential oil was poorly adsorbed onto the surface of hydroxyapatite nanoparticles.

Mentha species are one of the most used medical herbs due to their chemical constituents as menthol and menthone. Peppermint (*Mentha piperita*) essential oils are used as remedies in coughs, colds, mouth sinuses, pain relief and headaches, with properties related to antiviral activity against influenza, herpes and other viruses. The influence of peppermint essential oil (P-EO) on the surface of hydroxyapatite nanoparticles was studied [90] related to their morphological, physicochemical and antimicrobial properties. The results of the qualitative antimicrobial properties of P-EO and HAp-P are presented in **Table 1**.

Antimicrobial qualitative assay revealed that the peppermint had a significant inhibition effect on the microbial growth of the tested microorganisms, with the inhibition diameter ranging from 6 to 22 mm. The solvent DMSO did not affect the growth on solid media of any tested microbial strains. HAp had no inhibitory effect on the growth of the tested microorganisms and the most pronounced inhibition was observed in the case of *E. coli* tested strains in its two forms. The diameter of the inhibition growth area was 22 mm to 20 mm in the case of P-EO and a smaller inhibition zone of 8 mm for HAp-P. On *S. aureus* the inhibitory effect was related notably for P-EO and HAp-P with an inhibition zone in the range 7–8 mm. Essential oil extracted from peppermint contains active constituents that are responsible for eliminating bacterial pathogens, i.e. P-EO and HAp-P presented significant antibacterial activity with constituents acting on the cell membrane causing important morphological damage and destabilization of microbial membrane. Due to the worldwide emergence of *S. aureus* and *E. coli* strains which are resistant to conventional antibiotic therapy, there have been major concerns in public health area that conduced to the necessity of the developing of new antimicrobial compounds. Nano-sized powders of HAp doped with several metal ions that are known to possess antimicrobial properties as silver, zinc or cerium are used together with HAp in combination with essential oils. The effect of plants EOs and plants EOs-HAp combination regarding the antimicrobial activity is presented [91] in **Table 2** related to the diameter of inhibition zone-inhibition growth of tested bacterial strains.

The lavender EO inhibited the growth of all tested bacterial strains, as indicated the formation of inhibition zone ranged from 16 mm (*E. coli* ESBL 4493) to 24 mm (MRSA 1144). The HAp-L material was active against tested bacterial strains compared to HAp. The basil EO and HAp-B samples exhibited a lower inhibitory effect against the tested bacteria. On the other hand, HAp material had no effect on the growth of the selected bacteria.

The possibility of covering hydroxyapatite with different molecules, e.g. essential oils offer a solution to apply in food industry, in the idea that HAp is an essential component of human organism. In this regard, the potential use in medicine e.g. bone reconstruction could help the reducing of postoperative infections after different implants. In the case of hydroxyapatite nanotechnology have opened the gate to different applications in agriculture, food industry, medicine with the final target of improving human health and resistance to a continuous modification of pathogen agents.

Microbial Strains	P-EO	HAp-P	HAp	DMSO
Escherichia coli ATCC 25922	22 ± 0.2	8 ± 0.2	0 ± 0.1	0 ± 0.1
Escherichia coli C5	20 ± 0.3	7 ± 0.3	0 ± 0.1	0 ± 0.1
Pseudomonas aeruginosa ATCC 27853	10 ± 0.5	7 ± 0.5	0 ± 0.1	0 ± 0.1
Pseudomonas aeruginosa ATCC 9027	11 ± 0.3	6 ± 0.2	0 ± 0.1	0 ± 0.1
Staphylococcus aureus ATCC 25923	12 ± 0.3	10 ± 0.5	0 ± 0.1	0 ± 0.1
Staphylococcus aureus ATCC 6538	8 ± 0.2	7 ± 0.6	0 ± 0.1	0 ± 0.1
Methicillin-resistant Staphylococcus aureus (MRSA) 388	10 ± 0.5	0 ± 0.1	0 ± 0.1	0 ± 0.1
Enterococcus faecium DSM 13590	0 ± 0.1	0 ± 0.1	0 ± 0.1	0 ± 0.1
Candida parapsilosis ATCC 22019	0 ± 0.1	0 ± 0.1	0 ± 0.1	0 ± 0.1

Table 1.
The diameters of inhibition growth zones (mm).

Plant EOs and Plant EOs-HAp Combinations (Concentration)	Inhibition Zone (mm)					
	Lavander EO	HapL (10 mg/mL)	BasilEO	HApB (10 mg/mL)	Hap (10 mg/mL)	DMSO
Bacterial strain						
E. coli ATCC 25922	20 ± 1	15 ± 1	9 ± 2	7 ± 1	*-	*-
E. coli ESBL 4493	16 ± 0.5	10 ± 2	8 ± 1	6 ± 1	*-	*-
S. aureus 1426	25 ± 1	13 ± 2	10 ± 1	8 ± 1	*-	*-
MRSA 1144	24 ± 0.5	10 ± 2	7 ± 2	6 ± 1	*-	*-

* No inhibition zone observed.

Table 2.
Antimicrobial activities of plant EOs, HAp-B and HAp-L against Gram-positive and Gram-negative bacteria.

5. Conclusions

As it was presented, in the last few decades, nanotechnology reveals its benefit usage in different activity fields and in particular in biotechnology and agriculture. Fertilizer compounds are essential for our quality of soil and water for the development of plants in order to increase the crops in order to cover what is needed to sustain the food necessities all over the world. Therefore there exist a necessity to decrease nutrient casualties in fertilization, and to amplify the plant product by the operation of novel uses with assistance of nanotechnology and nanomaterials. This type of fertilization delivers the nutrients on request, control the use of chemical fertilizers that regulate growth and development of plants and raise the activity of target vegetal organism. In this regard have been presented the effects of different nanomaterials and nanoparticles upon selected plants culture as wheat, maize, soybean, etc. taking into account their growth morphological parameters and the nanoparticles influence upon plant metabolism. The present and future use of nanoparticles as micronutrients is affected by different risks related to nanotoxicity of micronutrients, a problem to be solved by an appropriate and safe circuit of nanoparticles in soil, water, plants and at last in human organism. In this regard, it is important to quantify nanomaterials concentration in water, soil and air, where the concentration of relevant nanomaterials is essential to define exposure, a problem to be solved by the modeling of environmental concentrations. Due to the rapid development of manufactured nanomaterials it is important to evaluate their environmental and health impact, and it was stated to assure a safe circuit from micronutrients used for plants crop increasing to the beneficiaries of these plants, i.e. animals and humans. The present work is an approximately extensive presentation of the present status of the application of nanomaterials and nanoparticles in agriculture i.e. in plants fertilization with accent to the plants growth parameters, possible toxic risks and application to the antimicrobial activity.

Acknowledgements

The authors thank for financial support to Romanian Ministry of Research and Innovation Contract No. 43PCCDI/2018 and Contract No. 23PCCDI/2018.

Author details

Daniela Predoi[1], Rodica V. Ghita[1*], Simona Liliana Iconaru[1], Carmen Laura Cimpeanu[2] and Stefania Mariana Raita[2]

1 National Institute of Materials Physics, Bucharest, Romania

2 University of Agronomic Sciences and Veterinary Medicine of Bucharest, Romania

*Address all correspondence to: ghitar@infim.ro

References

[1] Gogos A, Knauer K, Bucheli TD. Nanomaterials in plant protection and fertilization: current state foreseen application and research priorities. Journal of Agricultural and Food Chemistry. 2012;**60**(39):9781-9792

[2] Giraldo JP, Landry MP, Faltermeier SM, McNicholas TP, Iverson NM, Boghossian AA, et al. Plant nanobionics approach to argue photosynthesis and biochemical sensing. Nature Materials. 2014;**13**:400-408. DOI: 10.1038/nmat3890

[3] Talebi SM. Nanoparticle-induced morphological responses of roots and shoots of plants. In: Nanomaterials in Plants, Algae and Microorganisms. 2018. DOI: 10.1016/B978-0-12-811487-2.00006-2

[4] Miyazaki K, Islam N. Nanotechnology systems of innovation-An analysis of industry and academia research activities. Technovation. 2007;**27**:661-675

[5] Action Group on Erosion, Technology and Concentration (ETC group). Down on the Farm. Available from: http://www.etcgroup.org/upload/publication/80/02/etc_dotfarm2004.pdf

[6] Servin A, Elmer W, Mukherjee A, De la Torre-Roche R, Hamdi H, White JC, et al. A review of the use of engineered nanomaterials to suppress plan disease and enhance crop yield. Journal of Nanoparticle Research. 2015. DOI: 10.1007/s11051-015-2907-7

[7] Shew HD, Fichtner EJ, Benson DM. Aluminium and Plant Disease. St Paul: The American Phytopathological Society; 2007

[8] Navarro E et al. Environmental behavior and ecotoxicity of engineered nanoparticles to algae, plants and fungi. Ecotoxicity. 2008;**17**(5):372-386

[9] Tarafdar JC, Raliya R, Mahawar H, Rathore I. Development of zinc nanofertilizer to enhance crop production in pearl millet (*Pennisetum americanum*). Agricultural Research. 2014;**3**(3):257-262

[10] Khodakovskaya MV, Silva K, Biris AS, Dervishi E, Villagarcia H. Carbon nanotubes induce growth enhancement of tobacco cells. ACS Nano. 2012;**6**(3):2128-2135

[11] Khodakovskaya MV et al. Complex genetic, phothermal and photoacoustic analysis of nanoparticle-plant interactions. Proceedings of the National Academy of Sciences of the United States of America. 2011;**108**(3):1028-1033

[12] Alidoust D, Isoda A. Effect of gamma Fe_2O_3 nanoparticles on photosynthetic characteristics of soybean (*Glycine max* (L.) Merr): Foliar spray versus soil amendment. Acta Physiologiae Plantarum. 2013;**35**(12):3365-3375

[13] Pradhan S et al. Photochemical modulation of biosafe manganese naboparticles on *Vigna radiata* : A detailed molecular, biochemical and biophysical study. Environmental Science & Technology. 2013;**47**(22):13122-13131

[14] Linglan M, Chao L, Chunxiang Q, Sitao Y, Jie L, Fengqing G, et al. Rubisco activase mRNA expression in spinach: Modulation by nanoatanase treatment. Biological Trace Element Research. 2008;**122**(2):168-178

[15] Jo YK, Kim BH, Jung G. Antifungal activity of silver ions and nanoparticles against selected crop phytopathogenic fungi. Plant Disease. 2009;**93**(10):1037-1043

[16] He LL, Liu Y, Mustapha A, Lin MS. Antifungal activity of zinc oxide

nanoparticles against *Botrytis cinerea* and *Penicillium expansum*. Microbiological Research. 2011;**166**(3):207-215

[17] Cui H, Zhang P, Gu W, Jiang J. Application of anatase TiO_2 sol derived from peroxotitannic acid in crop diseases control and growth regulation. NSTI-Nanotech. 2009;**2**:286-289

[18] Jaberzadeh A, Moaveni P, Moghadam HRT, Zahedi H. Influence of bulk and nanoparticles titanium foliar application on some agronomic traits, seed gluten, and starch contents of wheat subjected to water deficit stress. Notulae Botanicae Horti Agrobotanici Cluj-Napoca. 2013;**41**(1):201-207

[19] Sims JT. Soil pH effects on the distribution and plant availability of manganese, copper and zinc. Soil Science Society of America Journal. 1986;**50**:367-373

[20] Yang K, Lin D, Xing B. Interactions of humic acid with nanosized inorganic oxides. Langmuir. 2009;**25**(6):3571-3576

[21] Eichert T, Kurtz A, Steiner U, Goldbach HE. Size exclusion limits and lateral heterogeneity of the stomatal foliar uptake pathway for aqueous solutes and water-suspended nanoparticles. Physiologia Plantarum. 2008;**134**(1):151-160

[22] Wang WN, Tarafdar JC, Biswas P. Nanoparticle synthesis and delivery by an aerosol route for watermelon plant foliar uptake. Journal of Nanoparticle Research. 2013;**15**(1):1-13

[23] Raliya R, Tarafdar JC. ZnO nanoparticle biosynthesis and its effect on phosphorous- mobilizing enzyme secretion and gum contents in Clusterbean (*Cyamopsis tetragonoloba* L). Agricultural Research. 2013;**2**(1):48-57

[24] Ahmad P, Rasool S. Emerging technologies and management of crop stress tolerance. In: Vol. 1. Biological Technique. San Diego: Academic Press; 2014

[25] Hong J et al. Evidence of translocation and physiological impacts of foliar applied CeO_2 nanoparticles on cucumber (*Cucumis sativus)* plants. Environmental Science & Technology. 2014;**48**(8):4376-4385

[26] Giannousi K, Avramidis I, Dendrinou-Samara C. Synthesis, characterization and evaluation of copper based nanoparticles as agrochemical against *Phytothphora infestans*. RSC Advances. 2013;**3**(44): 21743-21752

[27] Yin L et al. More than the ions: The effect of silver nanoparticles on *Lolium multiflorum*. Environmental Science & Technology. 2011;**45**(6):2360-2367

[28] Gogos A, Knauer K, Bucheli TD. Nanomaterials in plant protection and fertilization: Current state, foreseen applications, and research priorities. Journal of Agricultural and Food Chemistry;**60**(39):9781-9792

[29] Mueler NC, Nowack B. Exposure modeling of engineered nanoparticles in the environment. Environmental Science & Technology. 2008;**42**(12): 4447-4453

[30] Giessler-Blank S, Scheurmann R, Venzmer J, Lindsay D. Nanoemulsions and process for their preparation and their use as formulation of plant protection agents and/or pesticides and/ or cosmetic preparation. In: US Patent US20100041629. 2010

[31] Khan MR, Ahamad F, Rizvi TF. Effect of nanoparticles on plant pathogens. Advances in Phytonano-technology. DOI: 10.1016/ B978-0-12-815322-2.00009-2

[32] Khan MR, Rizvi TF. Nanotechnology: Scope and application

in plant disease management. Plant Pathology Journal. 2014;**13**:214-231

[33] Briat JF, Dubos C, Gaymard F. Iron nutrition, biomass production, and plant product quality. Trends in Plant Science. 2015;**20**:33-40. DOI: 10.1016/j.tplants.2014.07.005

[34] Marschner H, Rimmington GM. Mineral Nutrition of Higher Plants. SanDiego: Academic Press; 1995

[35] Dhoke SK, Mahajan P, Kamble R, Khanna A. Nanotechnology Development. 2013;**3**:e1-e5

[36] Sheykhbaglou R, Sedghi M, Shishevan T, Sharifi S. Notulae Scientia Biologicae. 2010;**2**:112-113

[37] Wang YX, Hussain SM, Krestin GP. European Radiology. 2001;**11**:2319-2331

[38] Liu XM, Zhang FD, Feg B, Zhang SQ, He XS, Wang RF, et al. Journal of Plant Nutrition and Fertilization. 2007;**11**:551-555

[39] Racuciu M, Creang DE. Journal of Magnetism and Magnetic Materials. 2007;**311**:291-294

[40] Shahrekizad M, Ahangar AG, Mir N. EDTA-coated Fe_3O_4 nanoparticles: A novel biocompatible fertilizer for improving agronomic traits of sun flower (*Helianthus annuus*). Journal of Nanostructures. 2015;**5**:117-127

[41] Cieschit MT, Polyakov AY, Lebedev VA, Volkov DS, Pankratov DA, Veligzhanin AA, et al. Eco-friendly iron-humic nanofertilizers synthesis for the prevention of iron chlorosis in soybean (*Glycine* max) grown in calcareous soil. Frontiers in Plant Science;**10**:Article 413. DOI: 10.3389/fpls.2019.00413

[42] Ali B, Gill RA, Yang S, Gill MB, Farooq MA, Liu D, et al. Regulation of cadmium-induced proteomic and metabolic changes by 5-aminolevulinic acid in leaves of *Brassica napus* L. PLoS One. 2015;**10**:1-23

[43] Usman M, Raheem Z, Ahsan T, Iqbal A, Sarfaraz ZN, Haq Z. Morphological, physiological and biochemical attributes as indicators for drought tolerance in rice (*Oryza sativa L.*). European Journal of Biological Sciences. 2013;**5**:23-28

[44] O'Conner D, Peng T, Zhang J, Tsang DC, Alessi DS, Shen Z, et al. Biochar application for the remediation of heavy metal polluted land: A review of *in situ* field trials. Science of the Total Environment. 2018;**619**:815-826

[45] Rizwan M, Ali S, Rehhman MZu, Andrees M, Arshad M, Qayyum MF, et al. Alleviation of cadmium accumulation in maize (*Zea mays* L.) by foliar spray of zinc oxide nanoparticles and biochar to contaminated soil. Environmental Pollution. 2019;**248**:358-367

[46] Pullagurala VLR, Adisa IO, RaWat S, Kalagara S, Hernandez-Viezcas JA, Peralta-Videa JR, et al. Finding the conditions for the beneficial use of ZnO nanoparticles towards plants—A review. Environmental Pollution. 2018;**241**:1175-1181

[47] Li C, Wang P, Lombi E, Cheng M, Tang C, Howard DL, et al. Absorption of foliar-applied Zn fertilizers by trichomes in soybean and tomato. Journal of Experimental Botany. 2018;**69**:2717-2729

[48] Rizwan M, Ali S, Rehman MZ, Maqbool A. A critical review on the effects of zinc at toxic levels of cadmium in plants. Environmental Science and Pollution Research. 2019. DOI: 10.1007/s11356-019-04174-6

[49] Lacerda JS, Martinez HE, Pedrosa AW, Clemente JM, Santos RH,

Oliveira GL, et al. Importance of zinc for Arabica coffee and its effects on the chemical composition of raw grain and beverage quality. Crop Science. 2018;**58**:1360-1370

[50] Martinez HEP, Zabini AV, Cruz CD, Pereira AA, Finger FL. Differential tolerance to zinc deficiency in coffee-plant progenies. Journal of Plant Nutrition. 2011;**34**:1654-1674

[51] Heinlaan M, Ivask A, Blinova I, Dubourguier HC, Kahru A. Toxicity of nanosized and bulk ZnO, CuO, and TiO_2 to bacteria *Vibrio fisheri* and crustaceans *Daphnia magna* and *Thamnocephalus platyurus*. Chemosphere. 2008;**71**:1308-1316

[52] Rossi L, Fedenia LN, Sharifan H, Ma X. Effects of foliar application of zinc sulfate and zinc nanoparticles in coffee (*Coffeea arabica* L) plants. Plant Physiology and Biochemistry. 2019;**135**:160-166

[53] Salama ZA, El-Fouly MM, Lazova G, Popova LP. Carboxylating enzymes and carbonic anhydrase functions were suppressed by zinc deficiency in maize and chickpea plants. Acta Physiologiae Plantarum. 2006;**28**:445-451

[54] Rout GR, Das P. Effect of metal toxicity on plant growth and metabolism: I. Zinc. In: Lichtfouse E, Navarrete M, Debaeke P, Veronique S, Alberola C, editors. Sustainable Agriculture. Netherlands, Dordrecht: Springer; 2009. pp. 873-884

[55] Davarpanah S, Tehranifar A, Davarynejad G, Abadia J, Khorasani R. Effects of foliar application of zinc and boron nano-fertilizer on pomegranate (*Punica granatum* cv. Ardestani) fruit yield and quality. Scientia Horticulturae. 2016;**210**:57-64

[56] White P, Broadley M. Calcium in plants. Annals of Botany. 2003;**92**:487-511

[57] Davarpanah S, Tehranifar A, Abadia J, Val J, Davarynejad G, Aran M, et al. Foliar calcium fertilization reduces fruit cracking in pomegranate (*Punica granatum* cv. Ardestani). Scientia Horticulturae. 2018;**230**:86-91

[58] Ye Y, Medina-Velo IA, Cota-Ruiz K, Moreo-OlivasF, Gardea-TorresdeyJL. Can abiotic stresses in plants be alleviated by manganese nanoparticles or compounds? Ecotoxicology and Environmental Safety. 2019;**184**:109671

[59] Pandya DH, Mer RK, Prajith PK, Pandey AN. Effect of salt stress and manganese supply on growth of barley seedlings. Journal of Plant Nutrition;**27**:1361-1379. DOI: 10.1081/PLN-200025835

[60] Singh U, Gardea-Torresdey J, Bindraban P, Adisa I, Elmer W, White J, et al. Effects of manganese nanoparticles exposure on nutrient acquisition in wheat (*Triticum aestivum* L). Agronomy. 2018;**8**:158

[61] Pradhan S, Patra P, Mitra S, Dey KK, Jain S, Sarkar S, et al. Manganese nanoparticles: Impact on non-modulated plant as a potent enhancer in nitrogen metabolism and toxicity study both *in vivo* and *in vitro*. Journal of Agricultural and Food Chemistry. 2014;**62**:8777-8785

[62] Panyuta O, Belava V, Fomaidi S, Kalinichenko O, Volkogon M, Taran N. The effect of pre-sowing seed treatment with metal nanoparticles on the formation of the defensive reaction of wheat seedlings infected with eyespot causal agent. Nanoscale Research Letters. 2016;**11**:1-5

[63] Nekrosova GF, Ushakova OS, Ermakov AE, Uimin MA, Byzov IV. Effects of copper (II) ions and copper oxide nanoparticles on *Elodea densa Planch*. Russian Journal of Ecology. 2011;**42**:458-463

[64] Shah V, Belozerova I. Influence of metal nanoparticles on the soil microbial community and germination of lettuce seeds. Water, Air and Soil Pollution. 2009;**197**:143-148

[65] Lee W, An Y, Yoon H, Kweon H. Toxicity and bioavailability of copper nanoparticles to the terrestrial plants mung bean (*Phaseolus radiates*) and wheat *(Triticum aestivum):* Plant agar test for water-insoluble nanoparticles. Environmental Toxicology and Chemistry. 2008;**27**:1915-1921

[66] Liu R, Lal R. Potentials of engineered nanoparticles as fertilizers for increasing agronomic productions. Science of the Total Environment. 2015;**514**:131-139

[67] Delfani M, Firouzabadi MB, Farrokhi N, Makarian H. Some physiological responses of black-eyed pea to iron and magnesium nanofertilizers. Communications in Soil Science and Plant Analysis. 2014;**45**:530-540

[68] Pallavi, Mehta CM, Srivastava R, Arora S, Sharma AK. Impact assessment of silver nanoparticles on plant growth and soil bacterial diversity. 3 Biotech. 2016;**6**(2):254

[69] Hansch M, Emmerling C. Effects of silver nanoparticles on the microbiota and enzyme activity in soil. Journal of Plant Nutrition and Soil Science. 2010;**173**:554-558

[70] Gruyer N, Dorais M, Bastien C, DassylvaN,Triffault-BouchetG.Interaction between silver nanoparticles and plant growth. Acta Horticulurae. 2014;**1037**:795

[71] Garcia-Gomez C, Fernandez MD. Impacts of metal oxide nanoparticles on seed germination, plant growth and development. In: Comprehensive Analytical Chemistry. Vol. 84. Elsevier; 2019. p. 76. ISSN 0166-526X

[72] Clement L, Hure C, Marmier N. Chemosphere. 2013;**90**:1083-1090

[73] Manojlovic MS, Loncaric Z, Cabilovski RR, Popovic B, Karalic K, Ivezic V, et al. Biofortification of wheat cultivars with selenium. Acta Agriculturae Scandinavica Section B-Soil and Plant Science. DOI: 10.1080/09064710.20191645204

[74] Lyons GH, Lewis J, Lorimer MF, Holloway RE, Brace DM, Stangoulis JC, et al. High-selenium wheat: Agronomic biofortification strategies to improve human nutrition. Journal of Food, Agriculture and Environment. 2004;**2**(1):171-178

[75] Siddiqui MH, Al-Whaibi MH, Faisal M, Al Sahli AA. Nano-silicon dioxide mitigates the adverse effects of salt stress on *Cucurbita pepo* L. Environmental Toxicology and Chemistry. 2014;**33**(11):2429-2437

[76] Haghighi M, Afifipour Z, Mozafarian M. The effect of N-Si on tomato seed germination under salinity levels. Journal of Biological & Environmental Sciences. 2012;**6**:87-90

[77] Li B, Tao G, Xie Y, Cai X. Physiological effects under the condition of spraying nano-SiO_2 onto the *Indocalamus barbatus* McClure leaves. Journal of Nanjing Forestry University. 2012;**36**:161-164

[78] Janmohammadi M, Sabaghnia N, Ahadnezhad A. Impact of silicon dioxide nanoparticles on seedling early growth of lentil (*Lens culinaris* medik) genotypes with various origins. Agriculture & Forestry. 2015;**61**(3):19-33

[79] Xie Y, Li B, Zhang Q, Zhang C, Lu K, Tao G, et al. Effects of nano-SiO_2 on photosynthetic characteristics of *Indocalamus barbatus*. Journal of Northeastern Forestry University. 2011;**39**:22-25

[80] Capaldi Arruda SC, Silva ALD, Galazzi RM, Azevedo RA, Zezi Arruda MA. Nanoparticles applied to plant science. A review. Talanta. 2015;**131**:693-705

[81] Hernandez-Viezcas JA, Castillo-Michel H, Servin AD, Peralta-Videa JR, Gardea-Torresdey JL. Journal of Chemical Engineering. 2011;**170**: 346-352

[82] Yang F, Liu C, Gao F, Xu M, Wu X, Zheng L, et al. Biological Trace Element Research. 2007;**119**:77-88

[83] Meng HY. Genotoxicity of hydrated sulfur dioxide on root tips of *Allium sativum* and *Viccia faba*. Mutation Research. 2003;**537**:109-114

[84] Vitoria AP, Lea PJ, Azevedo RA. Phytochemistry. 2001;**57**:701-710

[85] Servin AD, Morales MI, Castillo-Michel H, Hernandez-Viezcas JA, Munoz B, Zhao L, et al. Synchrotron verification of TiO_2 accumulation in cucumber fruit a possible pathway of TiO_2 nanoparticle transfer from soil into food chain. Environmental Science & Technology. 2013;**47**:11592-11598

[86] Kottegoda N, Munaweera I, Madusanka N, Karunaratne V. A green slow-release fertilizer composition based on urea-modified hydroxyapatite nanoparticles encapsulated wood. Current Science. 2011;**101**:73-78

[87] Liu R, Lal R. Synthetic apatite nanoparticles as a phosphorous fertilizer for soybean (*Glycine max*). Scientific Reports. 2014;**4**:5686

[88] Montalvo D, McLaughlin MJ, Degryse F. Efficacy of hydroxyapatite nanoparticles as phosphorous fertilizer in Andisols and Oxisols. Soil Science Society of America Journal. 2015;**79**:551-558

[89] Predoi D, Groza A, Iconaru SL, Predoi G, Barbuceanu F, Guegan R, et al. Properties of basil and lavender essential oils adsorbed on the surface of hydroxyapatite. Materials. 2018;**11**:652

[90] Badea ML, Iconaru SL, Groza A, Chifiriuc MC, Beuran M, Predoi D. Peppermint essential oil-doped hydroxyapatite nanoparticles with antimicrobial properties. Molecules. 2019;**24**:2169

[91] Predoi D, Iconaru SL, Buton N, Badea ML, Marutescu L. Antimicrobial activity of new materials based on lavender and basil essential oils and hydroxyapatite. Nanomaterials. 2018;**8**:291

Chapter 5

Nutritive Solutions Formulated from Organic Fertilizers

Juan Carlos Rodríguez Ortiz

Abstract

This chapter shows how organic fertilizers can provide essential nutrients soluble to plants, so as to be used in hydroponic systems in its various forms. Such materials are an important source of macro- and micronutrients. This form of plant nutrition can contribute to the sustainable production of food, both in developed and developing countries. Nutrient solutions can be formulated when soluble nutrients are extracted from the solid phase of organic manure. In some vegetables, equal yields, or sometimes higher, have been obtained in nutritive solutions formulated with synthetic chemical fertilizers. It has also been documented that the resulting edible products can be of a better nutraceutical quality. Ions can be obtained by means of preparations based on teas, extracts, leachates, digestate, urine, aquaculture, etc. Subsequently they must be diluted in water until reaching a level of electrical conductivity according to the tolerance levels of the crop to be established. The heterogeneity of the chemical composition of the solutions obtained is the main point that must be attended with the greatest possible precision to formulate the nutritive solutions and obtain satisfactory results. Therefore, it is necessary to measure the concentration of macro- and micronutrients (NO_3^-, NH_4^+, $SO_4^=$, $H_2PO_4^-$, K^+, Ca^{++}, Mg^{++}, Fe^{+++}, Cu^{++}, Mn^{++}, Zn^{++}, Cl^-) as well as the Na^+ ion (which is usually at high levels); it will also be necessary to adjust the pH. In addition, the chapter presents a broad overview and a series of research results in recent years: composition of solutions, nutrient supplements, substrates, and floating root trials in tomato, lettuce, cantaloupe melon, and green fodder. The environmental implications of inappropriate formulations are also analyzed. The nutritious solution, formulated from organic fertilizers, is not only an alternative for the nutrition of agricultural crops, but it also represents a more efficient way to use these resources.

Keywords: production systems, soilless, hydroponics, organic agriculture, plant nutrition

1. Introduction

The cyclical dynamics of the elements allow their reuse in ecosystems but also in agroecosystems. Organic matter represents a phase where they are partially and momentarily retained to follow the flow to various destinations, such as soil. Possible sources of nutrients, derived from reused or recycled materials, include wastewater; sewage sludge; biosolids; animal manure; urban waste; compost; vermicompost; digestate; biocarbon; inorganic by-products such as struvite, ammonium sulfate, and food waste; agribusinesses; and other industries [1].

This chapter focuses on manure, which is often the most available in the world's producing areas and is an important source of macro- and microelements for plant. For example, global manure nitrogen (N) production increased from 21.4 Tg N yr^{-1} in 1860 to 131.0 Tg N yr^{-1} in 2014, with a significant annual upward trend (0.7 Tg N yr^{-1}, $p < 0.01$), according to estimates of Zhang et al. (2017). These authors mention that cattle dominated the nitrogen production of manure and contributed 44% of total manure nitrogen production in 2014, followed by goats, sheep, porks, and poultries. The application of nitrogen from manure to farmland accounts for less than one-fifth of the total nitrogen production of manure during the study period.

Manure nitrogen production is expected to increase in the coming decades due to the growing demand for livestock populations as a result of increased human populations and changes in the structure of the diet with higher meat consumption (Herrero and Thornton, 2013).

While, in each country, there are significant resources of organic materials as sources of plant nutrients, their commercial use in hydroponics may be feasible if there is high availability and affordable costs, and on the other hand, they must be accompanied by guarantee of safety and food safety. This production technique is very promising for food production and efficient use of water and nutrients.

2. Formulation of the organic nutrient solution

The nutrient solution is a homogeneous mixture of water, ions (cations and anions), and oxygen that promote the growth and development of the vegetable species. Five steps are necessarily followed for the formulation of the ONS (**Figure 1**).

2.1 Step 1. Organic source selection

Oganic sources can be from different manures: bovine, poultry, sheep, goat, pork, etc. They must ensure the absence of microorganisms through effective composting and laboratory analysis to support it [2]. They must also have low heavy metal content, below the legal limits of each country. These requirements will be retaken in a space later.

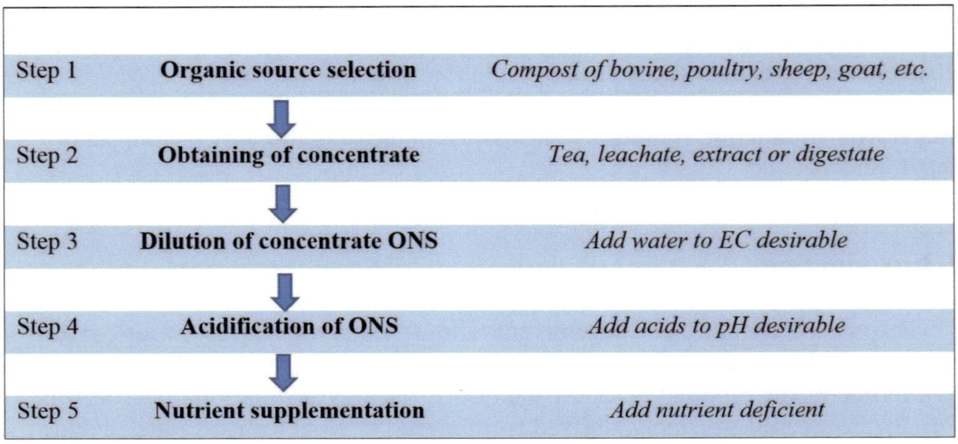

Figure 1.
Steps for the formulation of the ONS.

2.2 Step 2. Obtaining of concentrate

The concentrate is obtained from the solid organic materials, the main ones, which are the focus of this chapter, as follows: tea, leachate, extract, and digestate.

Compost tea: A "cold brewing" process, allowing growth of the organisms extracted from the compost [3].

Compost leachate: Water that drains, by oversaturation (excess moisture) of the material, during the composting process [4].

Extract: It is the product of passing water through the compost [4].

Digestate: Material remaining after various digestion processes have been applied to biomass or waste products such as animal manure, sewage sludge, and urban waste [1].

The concentrate can be obtained for unique extraction and sequential extraction.

2.3 Unique extraction

The ratios of solid and extracting organic material (usually water) are from 1:2 to 1:10; in a v:v ratio, rest times vary, typically from 8 to 48 h. The main parameter to measure is the electric conductivity (EC) of the solutions obtained and may vary due to the organic substrate, solid and extracting ratios, incubation time, and temperature of the solution, mainly (**Figure 2**). In 2013, González and colleagues studied the EC's relationship with the origin of vermicompost used in extraction (grass plus sheepman and more manure of sheep and cattle), the water/vermicompost ratio (1:2, 1:4, and 1:6), and the time (8, 16, and 24 h). They conclude that the origin of vermicompost has a high correlation with the EC, the ratio 1:2 (vermicompost/water) offers the advantage of obtaining concentrated teas with EC values, and the most suitable incubation time for tea extraction is 8 h.

Table 1 shows the total dissolved salts in a single extraction and **Table 2** for sequential extraction. It is observed that with sequential extraction it is possible to extract more dissolved solids than simple extraction, but more time is required.

2.4 Sequential extraction

Figure 3 shows the electrical conductivity of sequential extraction with poultry and bovine compost and water. The test was performed by mixing the compost with distilled water in a 1:2 (v/v) ratio with 48 h rest time between each extraction. The dynamics of the curve show that the soluble ions (measured by the EC) are released by describing a negative exponential function; the correlations had determination ratios of R^2 = 0.9388 and R^2 = 0.9042, in hen and bovine, respectively. The curves are stabilized

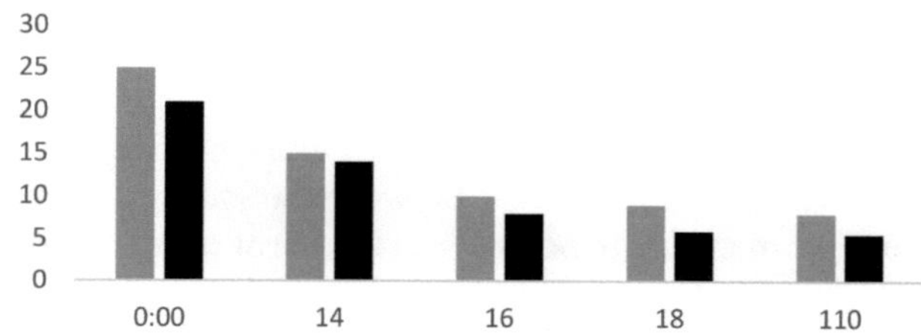

Figure 2.
EC that we obtained in five ratios dilution: 1:2, 1:4, 1:6, 1:8, and 1:10, with 24 h of rest in bovine and poultry compost extract.

Dilution ratio*	Rest time (h)	Fluid recovery (%)**	Volume recovery	Bovine	Poultry	Bovine	Poultry
						Dissolved salts	
				dS m^{-1}		§mg L^{-1}	
1:2	24	50	1	20.5	26.5	12.3	15.9
1:4	24	75	3	14	15	25.3	27
1:6	24	80	4.8	8	10	23	28.8
1:8	24	90	7.2	6	9	26	39
1:10	24	95	9.5	5.6	8	32	45.6
1:12	24	95	11.4	4	7	27.5	47.88
1.14	24	100	14	3	4.7	25.2	40
1:16	24	100	16	2.5	4	24	38.8

**Ratio solid material: volume of water applied.*
***Volume of water recovering from applied; the remaining percentage is retained by the solid phase.*
§The mg of salts dissolved per liter of liquid concentrate (factor 0.6 was used to convert from dS m^{-1} to mg L^{-1})

Table 1.
Content of dissolved salts extracted (a single extraction with 24 h rest).

Extraction	Dilution ratio*	ART§§	Fluid recovery (%)**	Volume recovery	Bovine	Poultry	Bovine	Poultry
							Dissolved salts	
					dS m^{-1}		§mg L^{-1}	
1	1:2	48	50	1	20.5	26.5	12.3	15.9
2	1:2	96	60	1.4	6.1	15.5	4.4	13
3	1:2	144	80	1.6	2.6	5.67	2.5	5.44
4w	1:2	192	85	1.7	1.6	3.3	1.6	3.4
5	1:2	240	90	1.8	1.2	1.83	1.3	2
6	1.2	288	95	1.9	1.2	1.2	1.35	1.35
7	1.2	336	100	2	1.2	1	1.44	1.2
8	1.2	384	100	2	1.2	0.9	1.44	1
						Σ	**26.33**	**43.3**

**Ratio solid material: volume of water applied*
***Volume of water recovering from applied; the remaining percentage is retained by the solid phase.*
§The mg of salts dissolved per liter of liquid concentrate (factor 0.6 was used to convert from dS m^{-1} to mg L^{-1})
§§Accumulated rest time

Table 2.
Salt content dissolved by sequential extraction (eight extractions in the same material with 48 h of rest between each extraction).

from the fifth extraction between 1.2 and 1.8 dS m^{-1} and continue with little variation until the eighth extraction. The ion balance, between the solid and aqueous phase of the mixtures, allows organic materials to be used as a source of nutrients for plants.

2.5 Dilution of concentrate to desirable EC

EC is generally used to indicate the total concentration of ionized constituents in water (Rodríguez et al., 2006). The concentrates shall be diluted with the irrigation water until the desired electrical conductivity is reached for the crop to be established (usually at 1–2 dS m^{-1}).

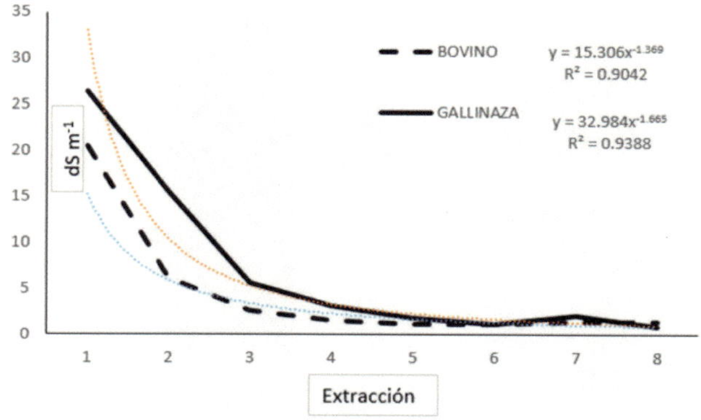

Figure 3.
Electrical conductivity of concentrates obtained sequentially from poultry and bovine compost.

2.6 Acidification of ONS

The pH indicates the degree of acidity or basicity of the solutions and is relevant by the availability of plant nutrients. **Figure 4** shows pH behavior in sequential extractions, in both composts (bovine and poultry). The pH range was 7–7.8, neutral to alkaline, indicating the possible presence of ions such as Ca^{2+}, Na^{+}, Mg^{2+}, HCO^{3-}, and CO_3^{2-}. The pH suitable for most plants in hydroponic systems is between 5 and 6 (Rodríguez et al., 2006), so organic nutrient solutions must be acidified that will partially eliminate carbonates and bicarbonates. Chemical or organic acids can be utilized; in the case of the test we conducted with bovine and poultry manure, the amount of sulfuric acid that we applied to lower pH from 7.4 to 6 per liter of ONS is 0.1 μL or 60 mL of acetic acid.

2.7 Nutrient supplementation

With the measurement of EC and pH, hydroponics solutions can be assessed. Up to this point, it is possible to use the nutritive solution obtained in small- and medium-sized plants, such as baby lettuces shown in **Figure 5**.

To produce higher biomass plants, for example, solanacea, cucurbitaceae, etc., it is recommended to have an analysis of the contents of essential elements in order to supplement in the organic nutrient solution, which can be very variable as shown in **Table 3**. The organic nutrient solutions are deficient in most essential elements when compared to known nutrient solutions. Therefore, it is necessary to supplement them with organic or inorganic sources, depending on the production system being worked.

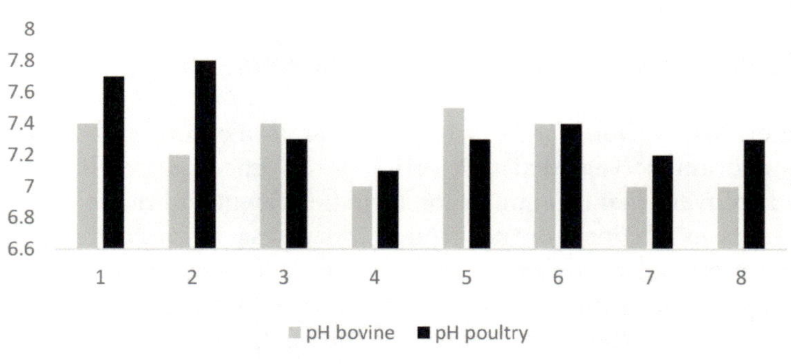

Figure 4.
pH of concentrates obtained sequentially from poultry and bovine compost.

Figure 5.
Twenty days after transplanting baby lettuce produced with chemical synthetic fertilizer (A) and manure extract (B). Both cultivated in floating root system with pH 6 and EC = 1.5 dS m^{-1}.

Author	1	2	3	4	Steiner solution	Sánchez
					References	
OM	VCG	VCP	P + B	VCB	—	—
Dilution	1:10	1:10	1:6	1:20	—	—
EC dS m^{-1}	2	2	4	1	2	2
		mg L^{-1}				
N	74.9	81.7	313.87	219	168	200
P	16.2	16.2	20.01	18.2	31	60
K	166.6	180.4	174.91	230	273	250
Ca	486	49	41.12	1.32	180	200
Mg	42.8	43.9	32.94	520	48	60
S	—		54.40	—	336	200
Fe	—	—	—	0.49	—	1
Cu	—	—	—	0.13	—	0.01
Mn	—	—	—	0.089	—	0.7
Zn	—	—	—	0.19	—	0.01

(1) Pant [5]; (2) Pant (2011); (3) González et al. [6]; (4) Ochoa et al. [7]. OM = organic material; VCG = vermicompost poultry manure; VCG = vermicompost poultry manure; CBv = compost bovine

Table 3.
Macro- and micronutrient content in some organic nutrient solutions.

3. Methods for the separation of ions in aqueous solution

In the literature, various methods for the separation of ions and cations (solute) in aqueous solution (solvent) exist. Selection of the process depends on the purity or degree of recovery that is required for both the solvent and the solute. Nowadays there are processes ranging from adsorption processes using activated carbon, which is one of the most economical materials, to reverse osmosis processes [8].

There are various chemical causes or reasons by which materials are related to ions or cations, among them attraction forces or electrostatic repulsions (mainly present in inorganic compounds), dispersive forces, or π-π interactions (organic compounds such as organic matter) stand out [9].

Among the materials used for the retention of inorganic compounds such as metals, metalloids, and heavy metals include activated carbons, zeolites, clays, lignocellulosic materials, carbon nanotubes, composites from green materials such as mixed cellulose with iron oxides and also can be used an ion exchange resins and membranes [10, 11]. In the case of organic compounds, the material traditionally used in Mexico and other countries is activated carbon.

A process that could be applied in the separation of ions and cations from organic fertilizers' derived mixture and whose main components are essentially potassium (K^+), nitrate (NO_3^-), and phosphate (PO_4^-) mixed with high organic matter content, which are an "interference" in the separation processes due to its high degree of complexity in the chemical structure, would be a sequential adsorption process [12, 13].

Therefore, as an ideal process for the elimination of this type of "interference" and the possible recovery of the ions and cations of interest, a cycle of separations must be done. First, the ion mixture must be placed in contact with carbon-based materials (this material already impregnated with that organic material can also be used for fertilizer) followed by cycles of adsorption columns with special ion exchange resins for each one of the ions. With this process, we can recover each of the components of the mixture and allocate them to the preparation of a nutrient solution according to each crop's needs.

4. Safety of organic materials

The main concern associated with the use of organic materials is mainly related to the possible presence of unwanted components, such as microbial pathogens, heavy metals, organic pollutants, waste pharmaceuticals, and personal care products, which threaten public health when undertreated. For example, organic materials could contain pesticide residues if obtained from some crop residues or antibiotics used in the diets of breeding animals, if excrement is used.

4.1 Heavy metals

The problem with regard to heavy metals is one of the most studied, and there is a vast literature dedicated to the subject. It is well known that concentrations of heavy metals above certain limits can lead to crop toxicity and may enter the food chain. The contents of MP in organic materials is very varied, since it depends on several factors, including the origin of the product, the feeding of livestock, etc. Rodriguez et al. [14] report the following total concentrations of heavy metals in cattle compost (in ppm): As 2.0 (−0.3), Cd 0.21 (−0.06), Hg <0.01, and Pb 5.9 (−1.01) and, for bovine lombricompost (in ppm), As 3.6 (−0.90), Cd 0.46 (−0.10), Hg <0.01, and Pb 16 (−2.60). For its part, Pane et al. [15] report the following heavy metal content in artichoke compost that was used to obtain nutrient solutions (78.0% artichoke, 20% woodchips, and 2% mature compost) (in ppm): Cd 0.38, Cr 20.69, Cu 21.01, Pb 13.45, Zn 13.45, and Zn 70.50, all below legal limits.

4.2 Pathogens

Depending on the source of the original material, the risks of contamination of unwanted organisms, such as pathogens, vary and are the highest in wastewater and excrement products.

Organic fertilizer production processes eliminate many pathogens as they include inactivation mechanisms such as very high temperatures, solar radiation, hydrolysis in strongly acidic or basic media, chemicals that affect pathogens, competition with other microorganisms, time, etc. (World Health Organization, 2018) [16]. If handled properly, composting can reduce pathogen levels [17]. In the inactivation of nonpathogenic *Escherichia coli*, pathogenic *E. coli* O157:H7, and *Salmonella* spp., several types of waste, such as animal manure and sewage sludge, have been reported during composting [18]. However, the persistence of *Listeria* spp., *Salmonella* spp., and nonpathogenic *E. coli* during composting [19] and the survival of *Salmonella* spp. and nonpathogenic *E. coli* in mature composts [20]. Most research on *E. coli* and *Salmonella* spp. have focused on manure or sewage sludge, but little attention has been paid to other substrates, such as green waste.

With regard to temperature, in many small composting units, degradation activity is limited by low temperature, well below 55°C. This is a very serious limitation when it comes to disinfection, since for many pathogens there is little or no reduction to temperatures below 50°C [16].

According to the US Environmental Protection Agency (US EPA) standard, Class A compost should not exceed the maximum *Salmonella* spp. limits (less than 3 most likely numbers [NMP]/4 g) or thermotolerant coliforms (less than 1000 NMP/g). The final amounts of bacteria, biological and viral, depend on the type of treatment used.

The current trend adopted in this field is to establish rigid rules that control the production process as well as to establish transport, packaging, and storage standards rather than setting pathogen limits on final products. For example, to acquire the characteristics necessary to be used in agriculture, sludge must undergo an additional disinfection process that ensures the reduction of the density of pathogens [16].

With regard to the risks of pathogens in organic fertilizers, it can be said that hazards can be excluded when production is industrialized, and this includes several disinfection procedures (pasteurization, drying, chemical media, etc.).

In addition, more or less stabilized organic substances, if poorly preserved and stored, can serve as excellent substrates for pathogens and become carriers of infections [21].

In the use of organic fertilizers, it is necessary to apply the precautionary principle, with the adoption of protective measures if there are suspicions that the products present a risk to public health or the environment. On the other hand, the danger of organic fertilizers and their amendments is certainly related to the end use of products.

Many organic compounds persist for long periods in soil, subsoil, aquifers, surface water, and aquatic sediments. These compounds, which can be of low or high molecular weight and that resist biodegradation, are known as recalcitrant. Many pesticides, mainly herbicides, have this characteristic [22].

Composting has been widely used for the remediation of organic pollutants as it, with adequate aeration, water, C-to-N ratio, and duration, accelerates their destruction [23]. The degradation of pesticides during composting depends on the pesticide and the substrate on which it is co-composted [24]. Strom [25] reported on the breakdown of organophosphorous pesticides and carbamates during composting. However, organochlorinated insecticides are resistant to degradation (Buyuksonmez et al., 1999). Differences in degradation may be related to inherent differences in the biological metabolism of the compound but may also be related to the composting process. Short-term composting (<60 days), which consists largely of the thermophilic phase, without adequate curing (mesophilic phase), may not be sufficient for the degradation of pesticides [26].

5. Humic acids, microorganisms, and hormones in organic materials

Organic materials, in addition to being a source of mineral elements (macronutrients and micronutrients), also provide the SN with other inseparable substances, among which are the microorganisms, humic acids (HA), and phytohormones.

5.1 Humic acids

Humic substances (HS) are the last substances resulting from chemical, biological, and physical transformations of plant and animal matter. The main compounds resulting from this transformation are humic acids, fulvic acids, and humines. Within these substances, humic acids, compounds soluble in alkaline solution and insoluble in acid solution and having a higher molecular weight, are the most important components [27, 28]. These substances, for their characteristics and effects on plants, have been considered as biostimulants [29].

HS are mineral compounds, among them essential elements for plants, mainly carbon, oxygen, hydrogen, nitrogen, sulfur, phosphorus (P), iron, copper, zinc and boron, in addition to functional groups among which stand out aromatic, aliphatic, carboxylic, and phenolic compounds (from [30–32]). HS are composed of hydrophobic fractions composed of aliphatic and aromatic compounds, while in another fraction, hydrophilic is composed of irregular humic fractions. These compounds, for their physicochemical characteristics, cause various effects on plants.

Among the metabolic processes that contribute to promote the growth and development of plants is the stimulation of the activity of key enzymes for the absorption and distribution of nutrients [33, 34]. The interaction of humic substances with proteins and lipids of the cell membrane improves the absorption of nutrition [35]. Mora et al. [36] mention that the presence of AH stimulated the activation of the H+-ATPase pump which led to a better distribution of NO_3^- from the root to the leaves. HSs can form latent complexes with metal ions, contributing to increased availability for root absorption as well as improving the distribution, within the plant, of metal ions [37].

There are various materials from which HS is obtained, which have been used in different crops in the hydroponic system. These substances have shown significant effects on these plants, improving growth and nutritional condition, mainly.

Haghighi and Teixeira [38] added 25 mg L^{-1} and 50 mg L^{-1} of HS extracted from forest soil moistified monthly to the nutrient solution used in the cultivation of tomato grown in perlite/vermiculite substrate. These HS were composed of 0.57% nitrogen, 0.03% phosphorus, and 4.5% potassium, with a pH of 4.5. Basically the addition of 50 mg L^{-1} of HS was the treatment that provoked the greatest effect in plants, increasing by 19% yield, 29% protein, 436% photosynthesis in growth stage, and 34% in fruiting stage. Other variables such as nitrate content, sugar content, and acidity in addition to antioxidant enzymes and chlorophyll were not affected by the presence of HS. These authors attributed the null effect on the abovementioned variables to the low concentrations of HS evaluated in the experiment.

Jannin et al. (2012) used 100 mg L^{-1} HS extracted from black peat for the formulation of Hoagland and Arnon nutrient solution (1950), for the cultivation of canola in floating root system. This material contained mainly 125, 40, 14, 9, and 2 mmol L^{-1} of potassium, sulfur, calcium (Ca), iron, and phosphorus, respectively, in addition to very low amounts of cytokinins such as zeatin, isopentenyladenine, and isopentenyladenosine. The plants were evaluated at days 1, 3, and 30 after the start of treatment, wherein the most significant effects were found at 30 days. The dry root weight was increased by 88%, while the total dry weight of the plant was increased by 29%. Nutrient absorption was increased with the presence of HS by 79% sulfur, 75% copper, 66% magnesium (Mg), 60% calcium, 57% nitrogen, and 47% potassium. Similarly, root nitrogen increased by 108% and sulfur increased by 76% in the leaf and 137% in the root. The abovementioned increases were the result of the expression of transporters present at the root responsible for the absorption of nitrogen and sulfur, in addition to the activity of the enzyme nitrate reductase.

The results showed that overall all materials were superior to the control. In particular 1 mg C L^{-1} increased the root length by 65% and the foliar area by 54%. The activity of the enzymes glutamine synthetase and glutamate synthetase, essential in nitrogen metabolism, were increased by 29% and 12%, respectively, with the addition of 10 mg C L^{-1}. Some important compounds in metabolism were increased. Protein content was increased by 43% in leaf and 8% in root at the concentration of 10 mg C L^{-1} and 1 mg C L^{-1}, respectively, while the foliar concentration of glucose and fructose were increased by 10% and 25% with the presence of 0.5 mg C L^{-1}. The activity of the enzyme phenylalanine ammonium lyase, participant in the production process of phenolic compounds, was increased by 51% by the presence of 1 mg C L^{-1}, so the content of phenolic compounds was increased by 15%.

5.2 Microorganisms and phytohormones

The use of nutritious solutions cast from organic fertilizers, such as composts, lombricomposts, vermicomposts, etc., may constitute an economic and environmental alternative to the use of chemical fertilizers for food production [39].

Organic fertilizers differ in quality, stability, and maturity because they depend on the organic waste and method by which they are prepared, so their chemical and biological composition varies and thus the nutritional composition and other elements that are present in the solutions obtained from them [40].

It is well documented that organic fertilizers contain soluble mineral nutrients such as nitrogen, phosphorus, potassium, magnesium, calcium, and other microelements, in addition to humic and fulvic acids, which the plant uses for its nutrition [39, 41]. But there is also the presence of phytohormones such as auxins, gibberellins, and cytokinins that are indispensable for the growth and development of plants [42–44].

In plants, phytohormones auxins, gibberellins, and cytokinins are the most common. Auxins, usually in the form of indolactic acid (AIA), are responsible for stimulating cell division, apical growth, and root branching [45]. Gibberellins, mainly in the form of gibberellic acid, are involved in various developmental and physiological processes, including seed germination, seedling emergence, stem and leaf growth, flowering, senescence, maturation of the plant [46]. Cytokinins play a key role in the process of cell division and bud growth and maintain photosynthetic activity and stoma opening during drought [47]. Therefore the presence of these hormones in organic fertilizers and the solutions obtained from them are of great importance and have to be considered; however, their presence has been less documented because they are difficult to detect and quantify, since they are usually found in trace concentrations and/or because they are immersed in a complex biological matrix, which makes their analysis quite difficult [44], but there are still some reports.

Zandonadi and collaborators reported the presence of indole-3-acetic acid (auxin) in humic acid extracted from a vermicompost. Zhang and collaborators (2014) reported the presence of cytokinins in tea also from a vermicompost. A study by Plant and collaborators (2012) reported the presence of isopentenyladenine-cytokinin, gibberellin 4 (GA4), and gibberellin 34 (GA34) in extracts of thermophilic compost based on chicken manure, waste vermicompost of food, and vermicompost based on chicken manure and the presence of gibberellin 24 (GA24) in vermicompost tea based on chicken manure. They also reported that a higher concentration of phytohormones can be attributed to increased activity of microorganisms present in fertilizers.

These phytohormones are produced by microorganisms present in organic fertilizers that come from soil and plant waste with which they are prepared [48, 49]. These microorganisms that produce these and other plant growth-promoting compounds are also known as plant growth-promoting microorganisms (PGPM) and are largely also responsible for biodegradation of the substrate or organic waste in the process of the production of organic fertilizers, mainly in composting [50], for example, *Azospirillum* spp. [51].

Among the microorganisms that produce auxins are those belonging to the genera *Azospirillum* spp. [52], *Azotobacter* spp. [53], *Rhizobium* spp. [54], *Bacillus subtilis* [55], *Bradyrhizobium* spp. [56], *Enterobacter* spp. [57], and *Trichoderma* spp. [58], to name a few. Within the production of gibberellins, *Azospirillum* spp. [59], *Bacillus* spp. [60], *Rhizobium* spp. [61], *Aspergillus* spp. [62], *Gibberella* spp. [63], and *Penicillium* spp. [64] are reported. The production of cytokinins is well characterized in microorganisms belonging to various genera such as *Azospirillum* [65], *Bacillus* spp. [66], and *Pseudomonas* spp. (Grokinsky et al., 2016) as well as the genera *Proteus*, *Klebsiella*, *Escherichia*, *and Xanthomonas* [43].

Although there is much research on the identification and quantification of phytohormones produced by various microorganisms (mainly bacteria and fungi that may be present in the organic waste and soil used for organic fertilizer processing and solutions obtained from them), studies related to the identification and quantification of phytohormones present in these are still scarce. This is due to the complexities necessary for the development of more sensitive and specific extractions, preparations and detection methods to analyze phytohormones. Quantification of phytohormones in organic waste solutions will be crucial for their complementation and supplementation with other compounds and improve food production more sustainably.

6. Trials of organic nutrient solutions in vegetables

6.1 Commercial and nutraceutical quality of compost extract in tomato fruits

We established a greenhouse trial with six treatments to determine the commercial and nutraceutical qualities and yield of tomato fruits (*Solanum lycopersicon*) fertilized with bovine compost and hen teas, and was treated with synthetic chemical fertilizers. Solutions were varied in electrical conductivity:

1. Compost extract of poultry manure with electric conductivity of 1.5 dS m^{-1}

2. Compost extract of poultry manure with electric conductivity of 2.0 dS m^{-1}

3. Compost extract of bovine manure with electric conductivity of 1.5 dS m^{-1}

4. Compost extract of bovine manure with electric conductivity of 2.0 dS m^{-1}

5. Steiner solution with electrical conductivity of 1.5 dS m^{-1}

6. Steiner solution with electrical conductivity 2.0 dS m^{-1}

Commercial materials were used as sources, which ensure the absence of pathogenic organisms. The cattle compost was the Organo Del brand (85% organic matter) and the hen the Meyfer brand, which has OMRI registration (37.7% organic matter). The extracts were prepared with a part compost and two water; the concentrate obtained was diluted with water until the indicated electrical conductivity and adjustment of pH to 6 with citric acid were obtained. The treatment of high-solubility synthetic chemical fertilizers used the Steiner solution.

The experiment was established in pots of 13 L capacity black plastic bags, and as a substrate was used, river sand (0.5–2 mm), previously sterilized. The genotype used was of habit determined variety "Caloro". The nutrient contents of the applied solutions, pH and EC, are presented in **Table 4**. All treatments had an average drainage of 20%. At 80 and 90–100 days after transplantation, the fruits with which the data were taken for evaluation were harvested.

The results indicate that treatments with organic solutions (hen and bovine extract) achieved production, quality in Brix grades, and phenol content statistically equal to those obtained in fertilizer, fertilizing treatments such as synthetic chemicals, regardless of the electrical conductivity of nutrient solutions (**Figure 6**). However, the antioxidant capacity was significantly higher in organic nutrient solutions with levels of 2 dS m^{-1} ($p < 0.05$).

6.2 Liquid digestate for hydroponic baby leaf lettuce (*Lactuca sativa* L.) cultivation

Ronga et al. [67] evaluated the effect of liquid digestate on the production of "baby" lettuce under hydroponic system over three cycles. This digestate was the product of anaerobic digestion of a mixture of corn, triticale, liquid dairy manure, and grape stems.

In the first and second cycle, the combination of perlite with standard nutrient solution (SNE), perlite with liquid digestate, solid digestate with SNE, solid digestate with liquid digestate, and soil control with SNE was evaluated. In the third cycle, the combinations were peat with SNE, peat with liquid digestate, pelletized digestate with SNE, and pelletized digestate with liquid digestate.

Chemical analyses showed that the liquid digestate contained 17% organic carbon, 0.34% nitrogen, and 0.95% potassium (K_2O) and has an electrical conductivity of 1.07 dS m^{-1} and a pH of 8.03, in addition to having the highest number of colony-forming units of all materials used (substrates and fertilizer materials) with 7.3 e+05 CFUs g^{-1}.

In the first cycle, treatments formed by the combination of solid digestate with SNE and perlite with liquid digestate produced higher dry weight of leaves, while the dry weight of root and total dry weight was benefited by the combination of perlite and digestate liquid. In addition, such treatment ensured the health of the crop by not finding coliforms in the plants.

Nutrient	Steiner solution	Steiner solution	Chicken manure tea	Chicken manure tea	Bovine compost tea	Bovine compost tea
	1.5	2	1.5	2	1.5	2
			dS m^{-1}			
N	115	153	39.2	49	28.45	35.56
P	23	31	9.2	11.5	8.15	10.18
K	207	277	107	133.75	103	128.75

Table 4.
N, P, and K composition of the treatments (mg L^{-1}).

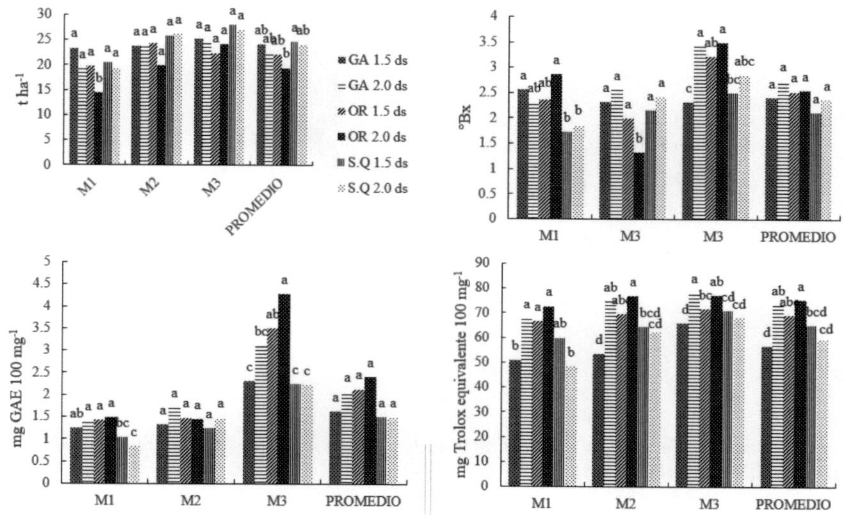

Figure 6.
Results of nutritious solution of hen, bovine, and chemical fertilizers: (a) yield by fruit cutting and average; (b) brix grades; (c) total phenols; and (d) antioxidant capacity.

In the second cycle, as in the previous cycle, the combination of solid digestate with SNE and perlite with liquid digestate produced greater dry weight of leaves. The same trend of the abovementioned variable was presented in the rest of the variables.

In the third cycle, the use of liquid digestate only equaled the SNE in the harvest index when the substrate was peat, while when the substrate was pelletized digestate, the liquid digestate produced higher plant height.

Based on the results shown, the authors consider the use of digestate for hydroponic production of lettuce to be a potential resource considering its low cost, environmental sustainability, agronomic interest, and microbial parameters.

6.3 Nutraceutical quality of cantaloupe melon fruits

The aim of the current study was to evaluate the nutraceutical quality of cantaloupe melon fruits fertilized with different organic fertilizer solutions (Preciado et al. (2015)); applied fertilization treatments consisted of an inorganic nutrient solution, compost tea, and vermicompost tea and leachate (leachate collected from vermicompost production) (**Figure 7**). The inorganic nutrient solution was prepared using highly soluble commercial fertilizers. The fertilizer solutions were adjusted to a pH of 5.5 and an EC of 2.0 dS m^{-1} via dilution with tap water to avoid phytotoxicity. The treatments were established in a completely randomized design using 10 plants per treatment, with each plant representing a treatment replicate.

The main conclusions of the present study are as follows. The applied nutrient solutions (compost tea, vermicompost tea and leachate, and inorganic Steiner solution) affected the nutraceutical quality of melon, as the fruits produced using the organic solutions exhibited higher antioxidant capacity and phenolic content than the chemically fertilized melons (**Figure 8**). It is feasible to recommend the application of vermicompost nutrient solutions (leachate and tea) as fertilizer alternatives for the production of hydroponic cantaloupe melon with an improved nutraceutical quality.

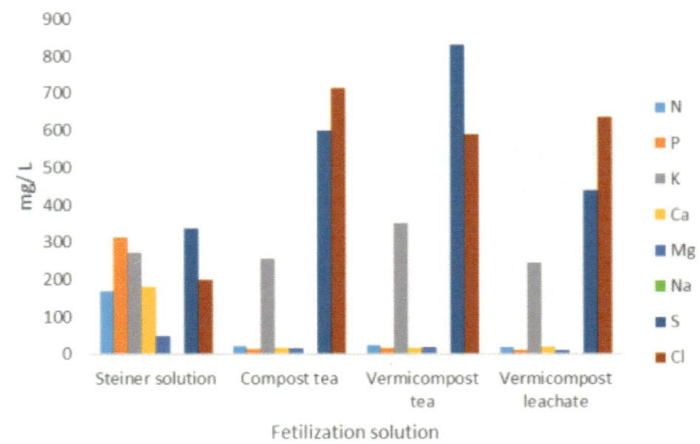

Figure 7.
Chemical composition of the nutrient solutions applied during the production of hydroponic cantaloupe melon in a greenhouse (Preciado et al., 2015).

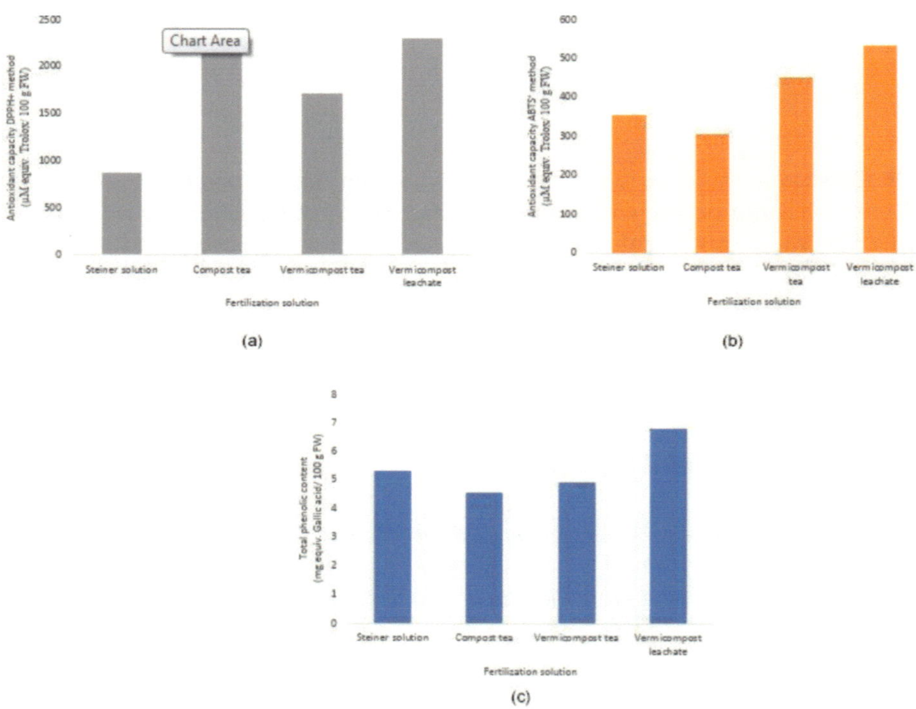

Figure 8.
Total phenolic content (a and b) and antioxidant capacity (c) of hydroponic cantaloupe melon fruits produced using different nutrient solutions.

6.4 Hydroponic green fodder

Salas et al. (2012) conducted a trial with the aim of evaluating the effect of organic nutrient solutions on yield, nutritional composition, total phenolic compounds, and in vitro antioxidant capacity of hydroponic green corn fodder produced in a greenhouse.

The treatments were vermicompost tea (TVC), compost tea (TC), and chemical solution (SQ) as a control and were applied from day 5 until harvest day. The concentration of nutrients in the treatments used is shown in **Figure 9**. Treatments were applied twice daily (8:00 and 19:00) on the aerial part of the fodder, with an average volume of 4.63 L^{-1} m^{-2} day^{-1}.

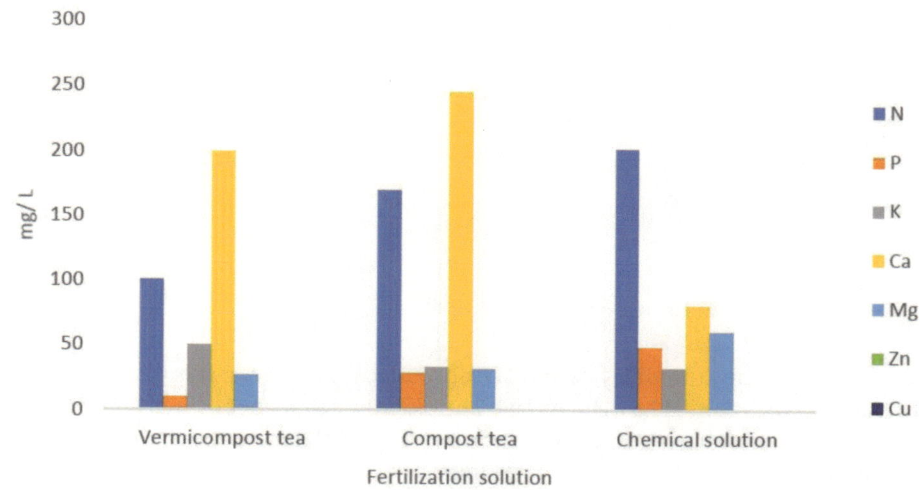

Figure 9.
Chemical composition of nutrient solutions applied in green fodder.

The yield, content of total phenolic compounds, and antioxidant capacity of the hydroponic green maize forage obtained were similar in organic and chemical fertilization treatments. Also, although differences in dry matter and protein content were found, all nutritional parameters evaluated were within the values reported as acceptable in good nutritional quality fodder (**Figures 10a–c** and **11**). On the other hand, the total phenolic content of organic and inorganically fertilized FVH was less than 1% dry base, so the consumption of such fodder does not pose health risks to livestock related to the consumption of these compounds. Therefore, it is advisable to use organic fertilization solutions in the production of fVH of maize in greenhouse, due to the advantages that such solutions would represent from the

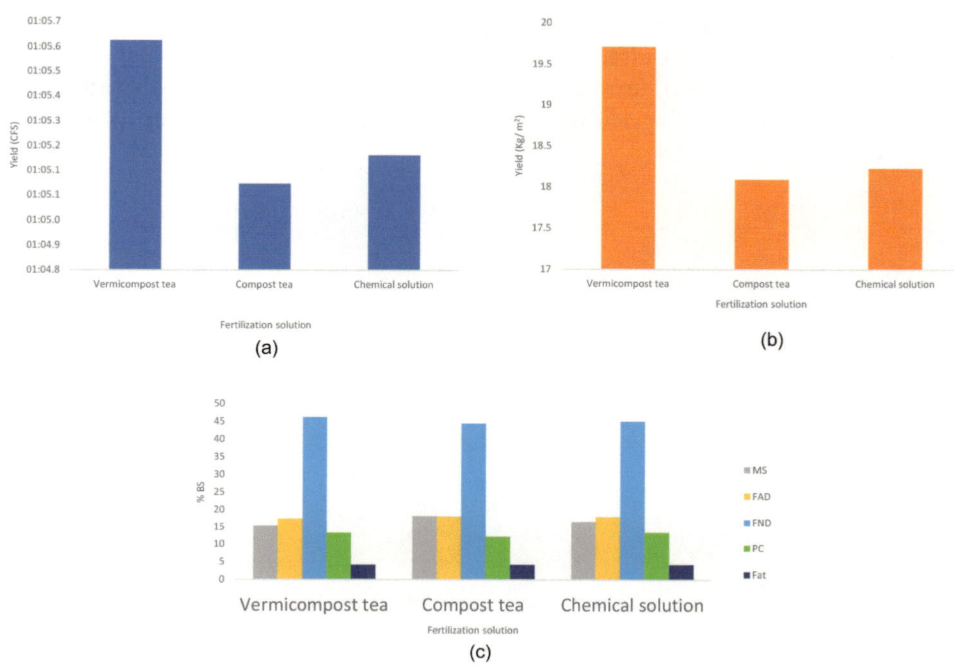

Figure 10.
Yield results and chemical composition of green fodder.

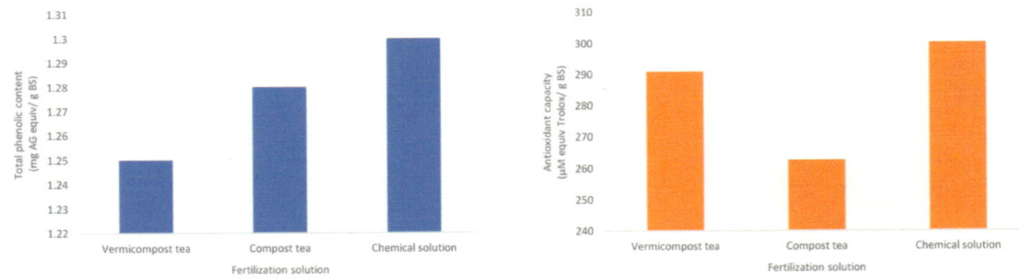

Figure 11.
Phenolic content and antioxidant content in green fodder.

point of view of sustainability by the use of available resources. It is recommended for future studies to evaluate the in vivo antioxidant properties of hydroponic green forage produced under organic fertilization as well as the identification of phenolic compounds contained in this type of fodder.

7. Conclusions

Organic fertilizers can provide essential nutrients soluble to plants, so as to be used in hydroponic systems in its various forms. Nutrient solutions can be formulated when soluble nutrients are extracted from the solid phase of organic manure, for this is essential to ensure that the organic materials used are harmless.

With these solutions it is possible to produce some vegetables without supplementing with other sources of nutrients (baby lettuce, chard, spinach, etc.). However, the solutions must be supplemented if solanaceas, cucurbits, or others plant groups are cultivated.

With organic solutions it is possible to have, in some vegetables, yields and commercial quality similar to solutions with chemical fertilizers. These vegetables also generally contain greater antioxidant capacity. The presence of other substances, in organic solutions, such as humic acids, phytohormones, and microorganisms, is responsible for the positive effects that have been obtained.

The nutritious solution, formulated from organic fertilizers, is not only an alternative for the nutrition of agricultural crops, but it also represents a more efficient way to use these resources.

Author details

Juan Carlos Rodríguez Ortiz
Facultad de Agronomía y Veterinaria UASLP, Mexico

*Address all correspondence to: juancarlos.rodriguez@uaslp.mx

References

[1] FAO. Código Internacional de Conducta Para el Uso y Manejo de Fertilizantes. C 2019/30. Roma. 2019. Available from: file:///C:/Users/jcrod/Desktop/Libro%20Hidroponia/JCR/FAO%202019.pdf

[2] Brinton W, Storms P, Evans E, Hill J. Compost teas: Microbial hygiene and quality in relation to method of preparation. Journal of Biodynamics. 2004;**2**:36-45

[3] Ingham ER. The Compost Tea Brewing Manual. 5a ed. Corvallis, Oregon: Soil FoodWeb Incorporated; 2005. 79 p

[4] Román Pilar María M, Pantoja MA. Manual de Compostaje del Agricultor Experiencias en América Latina. Santiago de Chile: Editado por FAO-ONU; 2013

[5] Pant AP, Radovich TJK, Hue NV, Talcottb ST, Krenekb KA. Vermicompost extracts influence growth, mineral nutrients, phytonutrients and antioxidant activity in pak choi (*Brassica rapa* cv. Bonsai, Chinensis group) grown under vermicompost and chemical fertilizer. Journal of the Science of Food and Agriculture. 2009;**89**:2383-2392

[6] González Solano K, De Las Nieves Rodríguez Mendoza M, Téllez LIT, Escudero JS, Cué JLG. Chemical properties of vermicompost "teas". Revista Mexicana de Ciencias Agrícolas Pub. Esp. Núm. 5 16 de mayo–29 de junio. 2013. pp. 901-911

[7] Ochoa-Martínez E, Figueroa-Viramontes U, Cano-Ríos P, Preciado-Rangel P, Moreno-Reséndez A, Rodríguez-Dimas N. Compost tea as organic fertilizer in the production of greenhouse tomato (*Lycopersicon esculentum* Mill.). Revista Chapingo Serie Horticultura. 2009;**15**(3):245-250

[8] Medellin-Castillo NA, Leyva-Ramos R, Padilla-Ortega E, Perez RO, Flores-Cano JV, Berber-Mendoza MS. Adsorption capacity of bone char for removing fluoride from water solution. Role of hydroxyapatite content, adsorption mechanism and competing anions. Journal of Industrial and Engineering Chemistry [Internet]. Elsevier BV; 2014;**20**(6):4014-4021. Available from: http://dx.doi.org/10.1016/j.jiec.2013.12.105

[9] Abdel Daiem MM, Rivera-Utrilla J, Sánchez-Polo M, Ocampo-Pérez R. Single, competitive, and dynamic adsorption on activated carbon of compounds used as plasticizers and herbicides. Science of the Total Environment [Internet]. Elsevier BV; 2015;**537**:335-342. Available from: http://dx.doi.org/10.1016/j.scitotenv.2015.07.131

[10] Liu L, Ji M, Wang F. Adsorption of nitrate onto $ZnCl_2$-modified coconut granular activated carbon: Kinetics, characteristics, and adsorption dynamics. Advances in Materials Science and Engineering [Internet]. Hindawi Limited; 2018;**2018**:1-12. Available from: http://dx.doi.org/10.1155/2018/1939032

[11] Shahid MK, Kim Y, Choi Y-G. Magnetite synthesis using iron oxide waste and its application for phosphate adsorption with column and batch reactors. Chemical Engineering Research and Design [Internet]. Elsevier BV; 2019;**148**:169-179. Available from: http://dx.doi.org/10.1016/j.cherd.2019.06.001

[12] Jamil S, Loganathan P, Listowski A, Kandasamy J, Khourshed C, Vigneswaran S. Simultaneous removal of natural organic matter and micro-organic pollutants from reverse osmosis concentrate using granular activated carbon. Water Research [Internet].

Elsevier BV; 2019;**155**:106-114. Available from: http://dx.doi.org/10.1016/j.watres.2019.02.016

[13] Zhang Y, Wang X, Jia H, Fu B, Xu R, Fu Q. Algal fouling and extracellular organic matter removal in powdered activated carbon-submerged hollow fiber ultrafiltration membrane systems. Science of the Total Environment [Internet]. Elsevier BV; 2019;**671**: 351-361. Available from: http://dx.doi.org/10.1016/j.scitotenv.2019.03.371

[14] Rodríguez O, Carlos J, Jáuregui JAA, Montoya AH, Fuentes HR, Ruiz FH, et al. Trace elements in fertilizers and manure used in organic and conventional agricultura. Revista Mexicana de Ciencias Agrícolas. 2014;**5**(4):695-701

[15] Pane C, Palese AM, Celano G, Zaccardelli M. Effects of compost tea treatments on productivity of lettuce and kohlrabi systems under organic cropping management. Italian Journal of Agronomy [Internet]. PAGEPress Publications; 2014;**9**(3):153. Available from: http://dx.doi.org/10.4081/ija.2014.596

[16] Vinnerås B, Agostini F, Jönsson H. Sanitation by composting. In: Microbes at Work. Berlin, Heidelberg: Springer; 2010. pp. 171-191

[17] Ryckeboer J, Mergaert J, Coosemans J, Deprins K, Swings J. Microbiological aspects of biowaste during composting in a monitored compost bin. Journal of Applied Microbiology. 2003;**94**:127-137

[18] Lung AJ, Lin CM, Kim JM, Marshall MR, Nordstedt R, Thompson NP, et al. Destruction of *Escherichia coli* O157:H7 and *Salmonella enteritidis* in cow manure composting. Journal of Food Protection. 2001;**64**(9): 1309-1314

[19] Droffner ML, Brinton WF, Evans E. Evidence for the prominence of well characterized mesophilic bacteria in thermophilic (50-70°C) composting environments. Biomass and Bioenergy. 1995;**8**:191-195

[20] Sidhu J, Gibbs RA, Ho GE, Unkovich I. Selection of *Salmonella typhimurium* as an indicator for pathogen regrowth potential in composted biosolids. Letters in Applied Microbiology. 1999;**29**:303-307

[21] Eregno F, Moges M, Heistad A. Treated greywater reuse for hydroponic lettuce production in a green wall system: Quantitative health risk assessment. Water. 2017;**9**(7):454

[22] Navarro S, Vela N, Navarro G. An overview on the environmental behaviour of pesticide residues in soils. Spanish Journal of Agricultural Research. 2007;**3**:357-375

[23] Taiwo AM. Composting as a sustainable waste management technique in developing countries. Journal of Environmental Science and Technology. 2011;**4**(2):93-102

[24] Barker AV, Bryson GM. Bioremediation of heavy metals and organic toxicants by composting. The Scientific World Journal. 2002;**2**: 407-420

[25] Strom PF. Evaluating pesticide residues in yard trimmings compost. BioCycle (USA). 1998;**39**(11):80

[26] Blewett TC, Roberts DW, Brinton WF. Phytotoxicity factors and herbicide contamination in relation to compost quality management practices. Renewable Agriculture and Food Systems. 2005;**20**(2):67-72

[27] Canellas LP, Olivares FL, Aguiar NO, Jones DL, Nebbioso A, Mazzei P, et al. Humic and fulvic acids as biostimulants in horticulture. Scientia Horticulturae. 2015;**196**:15-27

[28] Nardi S, Pizzeghello D, Schiavon M, Ertani A. Plant biostimulants:

Physiological responses induced by protein hydrolyzed-based products and humic substances in plant metabolism. Scientia Agricola. 2016;**73**(1):18-23

[29] Yakhin OI, Lubyanov AA, Yakhin IA, Brown PH. Biostimulants in plant science: A global perspective. Frontiers in Plant Science [Internet]. Frontiers Media SA; 2017;7. Available from: http://dx.doi.org/10.3389/fpls.2016.02049

[30] de Melo BAG, Motta FL, Santana MHA. Humic acids: Structural properties and multiple functionalities for novel technological developments. Materials Science and Engineering: C. 2016;**62**:967-974

[31] García AC, De Souza LGA, Pereira MG, Castro RN, García-Mina JM, Zonta E, et al. Structure-property-function relationship in humic substances to explain the biological activity in plants. Scientific Reports. 2016;**6**:20798

[32] Jindo K, Martim SA, Navarro EC, Pérez-Alfocea F, Hernandez T, Garcia C, et al. Root growth promotion by humic acids from composted and non-composted urban organic wastes. Plant and Soil. 2012;**353**(1-2):209-220

[33] Olaetxea M, Mora V, García AC, Santos LA, Baigorri R, Fuentes M, et al. Root-shoot signaling crosstalk involved in the shoot growth promoting action of rhizospheric humic acids. Plant Signaling & Behavior. 2016;**11**(4):e1161878

[34] Veobides-Amador H, Guridi-Izquierdo F, Vázquez-Padrón V. Las sustancias húmicas Como bioestimulantes de plantas bajo condiciones de estrés ambiental. Cultivos Tropicales. 2018;**39**(4):102-109

[35] Van Oosten MJ, Pepe O, De Pascale S, Silletti S, Maggio A. The role of biostimulants and bioeffectors as alleviators of abiotic stress in crop plants. Chemical and Biological Technologies in Agriculture. 2017;**4**(1):5

[36] Mora V, Bacaicoa E, Zamarreno AM, Aguirre E, Garnica M, Fuentes M, et al. Action of humic acid on promotion of cucumber shoot growth involves nitrate-related changes associated with the root-to-shoot distribution of cytokinins, polyamines and mineral nutrients. Journal of Plant Physiology. 2010;**167**(8):633-642

[37] Rasheed SM, Abdullah HM, Ali ST. Response of two hybrids of tomato (*Lycupersicun esculentum* Mill) to four concentration of humic acid fertilizers in plastic house condition. Journal of Tikrit University for Agriculture Sciences. 2017;**17**(1):1-12

[38] Haghighi M, Teixeira Da Silva JA. Amendment of hydroponic nutrient solution with humic acid and glutamic acid in tomato (*Lycopersicon esculentum* Mill.) culture. Soil Science and Plant Nutrition. 2013;**59**(4):642-648

[39] Maheshwari DK. Composting for Sustainable Agriculture. Cham, Switzerland: Springer; 2014. pp. 199-207

[40] Azim K, Soudi B, Boukhari S, Perissol C, Roussos S, Thami Alami I. Composting parameters and compost quality: A literature review. Organic Agriculture [Internet]. Springer Science and Business Media LLC; 2017;**8**(2): 141-158. Available from: http://dx.doi.org/10.1007/s13165-017-0180-z

[41] Guo X, Liu H, Wu S. Humic substances developed during organic waste composting: Formation mechanisms, structural properties, and agronomic functions. Science of the Total Environment [Internet]. Elsevier BV; 2019;**662**:501-510. Available from: http://dx.doi.org/10.1016/j.scitotenv.2019.01.137

[42] Arancon N, Edwards CA, Webster KA, Buckerfield JC.

In: Edwards CA, Arancon NQ, Sherman RL, editors. Vermiculture Technology: Earthworms, Organic Wastes, and Environmental Management. Boca Raton, Florida, USA: CRC Press; 2010

[43] Maheshwari DK, Dheeman S, Agarwal M. Phytohormone-producing PGPR for sustainable agriculture. In: Bacterial Metabolites in Sustainable Agroecosystem. Cham: Springer; 2015. pp. 159-182

[44] Zhang H, Tan SN, Teo CH, Yew YR, Ge L, Chen X, et al. Analysis of phytohormones in vermicompost using a novel combinative sample preparation strategy of ultrasound-assisted extraction and solid-phase extraction coupled with liquid chromatography–tandem mass spectrometry. Talanta. 2015;**139**:189

[45] Li S-B, Xie Z-Z, Hu C-G, Zhang J-Z. A review of auxin response factors (ARFs) in plants. Frontiers in Plant Science [Internet]. Frontiers Media SA; 2016;7:47. Available from: http://dx.doi.org/10.3389/fpls.2016.00047

[46] Camara MC, Vandenberghe LP, Rodrigues C, de Oliveira J, Faulds C, Bertrand E, et al. Current advances in gibberellic acid (GA 3) production, patented technologies and potential applications. Planta. 2018;**248**(5): 1049-1062

[47] Hwang I, Sheen J, Müller B. Cytokinin signaling networks. Annual Review of Plant Biology. 2012;**63**: 353-380

[48] Berg G. Plant–microbe interactions promoting plant growth and health: Perspectives for controlled use of microorganisms in agriculture. Applied Microbiology and Biotechnology. 2009;**84**(1):11-18

[49] Pii Y, Mimmo T, Tomasi N, Terzano R, Cesco S, Crecchio C. Microbial interactions in the rhizosphere: Beneficial influences of plant growth-promoting rhizobacteria on nutrient acquisition process. A review. Biology and Fertility of Soils. 2015;**51**(4):403-415

[50] Nozhevnikova AN, Mironov VV, Botchkova EA, Litti YV, Russkova YI. Composition of a microbial community at different stages of composting and the prospects for compost production from municipal organic waste. Applied Biochemistry and Microbiology. 2019;**55**(3):199-208

[51] Cassán F, Vanderleyden J, Spaepen S. Physiological and agronomical aspects of phytohormone production by model plant-growth promoting rhizobacteria (PGPR) belonging to the genus Azospirillum. Journal of Plant Growth Regulation. 2014;**33**(2):440-459

[52] Mehnaz S. Azospirillum: A biofertilizer for every crop. In: Plant Microbes Symbiosis: Applied Facets. New Delhi: Springer; 2015. pp. 297-314

[53] Husen E. Screening of soil bacteria for plant growth promotion activities in vitro. Indonesian Journal of Agricultural Science. 2016;**4**(1):27-31

[54] Jiménez-Gómez A, Menéndez E, Flores-Félix JD, García-Fraile P, Mateos PF, Rivas R. Effective colonization of spinach root surface by rhizobium. In: Biological Nitrogen Fixation and Beneficial Plant-Microbe Interaction. Cham: Springer; 2016. pp. 109-122

[55] Arkhipova T, Galimsyanova N, Kuzmina L, Vysotskaya L, Sidorova L, Gabbasova I, et al. Effect of seed bacterization with plant growth-promoting bacteria on wheat productivity and phosphorus mobility in the rhizosphere. Plant, Soil and Environment [Internet]. Czech Academy of Agricultural Sciences; 2019;**65**(6):313-319.

Available from: http://dx.doi.org/10.17221/752/2018-pse

[56] Subramanian P, Kim K, Krishnamoorthy R, Sundaram S, Sa T. Endophytic bacteria improve nodule function and plant nitrogen in soybean on co-inoculation with *Bradyrhizobium japonicum* MN110. Plant Growth Regulation. 2015;**76**(3):327-332

[57] Dhungana SA, Itoh K. Effects of co-inoculation of Indole-3-acetic acid-producing and-degrading bacterial endophytes on plant growth. Horticulturae. 2019;**5**(1):17

[58] Nieto-Jacobo MF, Steyaert JM, Salazar-Badillo FB, Nguyen DV, Rostás M, Braithwaite M, et al. Environmental growth conditions of Trichoderma spp. affects indole acetic acid derivatives, volatile organic compounds, and plant growth promotion. Frontiers in Plant Science. 2017;**8**:102

[59] Cassán F, Diaz-Zorita M. Azospirillum sp. in current agriculture: From the laboratory to the field. Soil Biology and Biochemistry. 2016;**103**: 117-130

[60] Tahir HA, Gu Q, Wu H, Raza W, Hanif A, Wu L, et al. Plant growth promotion by volatile organic compounds produced by *Bacillus subtilis* SYST2. Frontiers in Microbiology. 2017;**8**:171

[61] Nett RS, Contreras T, Peters RJ. Characterization of CYP115 as a gibberellin 3-oxidase indicates that certain rhizobia can produce bioactive gibberellin A4. ACS Chemical Biology. 2017;**12**(4):912-917

[62] You YH, Park JM, Kang SM, Park JH, Lee IJ, Kim JG. Plant growth promotion and gibberellin A3 production by *Aspergillus flavus* Y2H001. The Korean Journal of Mycology. 2015;**43**(3):200-205

[63] Salazar-Cerezo S, Martínez-Montiel N, García-Sánchez J, Pérez-y-Terrón R, Martínez-Contreras RD. Gibberellin biosynthesis and metabolism: A convergent route for plants, fungi and bacteria. Microbiological Research. 2018;**208**:85-98

[64] Khan AL, Waqas M, Lee IJ. Resilience of Penicillium resedanum LK6 and exogenous gibberellin in improving *Capsicum annuum* growth under abiotic stresses. Journal of Plant Research. 2015;**128**(2):259-268

[65] Castillo P, Molina R, Andrade A, Vigliocco A, Alemano S, Cassán FD. Phytohormones and other plant growth regulators produced by PGPR: The genus Azospirillum. In: Handbook for Azospirillum. Cham: Springer; 2015. pp. 115-138

[66] Bacon CW, Palencia ER, Hinton DM. Abiotic and biotic plant stress-tolerant and beneficial secondary metabolites produced by endophytic *Bacillus* species. Plant Microbes Symbiosis: Applied Facets [Internet]. India: Springer; 2014:163-177. Available from: http://dx.doi.org/10.1007/978-81-322-2068-8_8

[67] Ronga D, Setti L, Salvarani C, De Leo R, Bedin E, Pulvirenti A, et al. Effects of solid and liquid digestate for hydroponic baby leaf lettuce (*Lactuca sativa* L.) cultivation. Scientia Horticulturae. 2019;**244**:172-181

Chapter 6

Installation of Vegetable Based Roof Gardens in Schools From Recyclable Materials: A Study

Adriana Maria dos Santos, Mariana Paiva Baracuhy, Dermeval Araújo Furtado, Romulo Wilker Neri de Andrade, Jackson Rômulo de Sousa Leite and Fabiana Terezinha Leal de Morais

Abstract

The study aimed to reflect on the socio-environmental issues and the action of the gardens in urban/school spaces, considering garden as a methodological instrument for the interdisciplinary activities related to family farming, using the descriptive methodology and study of literary review with proposals of gardens using recyclable materials depicted through images created using the software AutoCAD. Through the study, it was possible to plan gardens using recyclable materials in environments of small spaces. The crops employed will be vegetables for school meals. The activities carried out in the garden contribute to the change in the habits and attitudes of students regarding the perception they possess of nature, the formation of awareness of respect and care, the need to conserve the environment and stimulate the pursuit of improvement of quality of life in other ways of seeing the activities performed by their own parents in the field.

Keywords: sustainability, waste, agriculture

1. Introduction

In view of the constant evolutions that technology imposes on agriculture and food production, especially in family farming associated with climatic phenomena, the uncertainties of an economically, socially, politically and technologically correct agriculture, as well as the absence of agricultural practices in the experience of young people in rural communities, raises a concern on the future of world agriculture and food production [1], with similar concerns occurring in Brazil.

In Nigeria, the idea that agricultural activity, especially in rural areas, is undervalued and provides few benefits for its practitioners [2], prevails on the part of young people.

Anjos and Caldas [3] cite that there is a very negative view populating the imagery of rural youths, a fact that, ultimately, reproduces the dominant stigma that rural spaces represent the place of "non-development", of the archaic, of the traditional.

According to Guthman [4] for the production of food, students, in the exercise of citizenship or as future agricultural workers, will be better able to understand the debates and controversies that underlie the production, creation and marketing of agricultural products, recognizing the limits and possibilities of models, both of intensive production and alternative models, these little valued and disseminated (family farming, agroforestry production, etc.), as well as recognizing the various aspects (environmental, scientific, political, economic, cultural, etc.) present in the different models of food production and understand the different tools of flexible teaching and learning, based on permaculture, that the gardens School (Gardens) can offer.

One of the alternatives to raise the perception about activities in the agricultural environment and the care of the environment is the use of a school garden, which can serve as a source of food and didactic activities, offering advantages to the communities involved, such as obtaining quality food at low cost and involvement in food and health programs developed by schools [5], contributing also to the knowledge of the 3 R's (reduce, reuse and recycle), integration of the community school in the performance of socio-environmental activities, encouraging the consumption of organic foods, providing students with experiences of agroecological practices for food production, so that they can be transmitted to their relatives and, consequently, apply them to home or community gardens [6].

In this context, the research aims to reflect on the environmental issues and the action of the gardens in urban/school spaces, taking allowance from the garden as a methodological instrument the interdisciplinary practices of activities related to agriculture family.

2. Materials and methods

The literature review presented below contains a synthesis of the latest studies on the production of vegetable gardens in urban areas, highlighting the production of vegetable gardens in schools. The methods used were studies of free area in school spaces of three schools of early childhood education, investigation of the needs and desires of the school community and researches on types and forms for plant production in urban areas and employability of recyclable materials in its construction. The software Auto Card, a tool for architectural drawings, enabling the creation of gardens for each space studied was used.

For the preparation of the gardens passive recycling materials may be used, through characterization of solid residues, materials that has been discarded by the local population, without appropriate destination for the environment, for example,

- Tires: used in the garden site
- Pet bottles: it is used to demarcate the Mandala (vegetable garden en circle) garden site and store rainwater for irrigation through the drip.
- Paper: fertilizer and base of the flowerbeds.
- Organic residues of food production in school: fertilizer.
- Gray water from the production of school feeding: irrigation and fertilization.
- Demolition wood: to assemble the structure of the vegetable gardens.

DOI: http://dx.doi.org/10.5772/intechopen.90721

2.1 Results and discussions

2.1.1 Urban agriculture

Urban agriculture is an activity that has been growing in Brazil and worldwide, according to FAO—Food and Agriculture Organization of the United Nations. This activity refers to the use of surfaces located in urban areas or in their respective peripheries for agricultural production and the creation of small animals intended for own consumption or for sale in local markets.

Some of the concepts about urban agriculture in general address their relationship with localization. For Dimoud and Nikolopoulou [7], the definition of urban agriculture refers to the location of the spaces within and around cities or urban areas. Therefore, the intra area refers to all spaces within cities that may have some type of agricultural activity, which can be individual or collective, in addition to being located in private or public areas such as squares or idle areas.

Wong [8] stated that the concept of urban agriculture goes beyond what is defined by the area of localization, which is therefore an interaction between the ecological and urban economic system, not being reduced only to the urban location.

Dimoud and Nikolopoulou [7] stated that this integration is made possible by the fact that urban agriculture has a set of activities (cultivation, breeding, fishing, etc.) that develop in the interior (Intraurban) or in the periphery (periurban) of the cities.

The development of urban or periurban agriculture is directly linked to the demographic and economic growth of cities, contributing to the reconfiguration of urban spaces through land use, population structures, social practices, among other factors.

The advantages of urban agriculture, includes the local development through the rational use of spaces, food security, formation of microclimates, maintenance of biodiversity, water drainage, harvesting of rain water, decreased temperature and income generation. Most common examples of urban agriculture are the community gardens that are most often installed in urban idle areas, which may be public or private, intended for cultivations of vegetables, medicinal plants, legumes, fruits and other foods, providing food for families living near these areas, or seedling production [9].

Urban gardens have differentiated configurations, where they almost always correspond to the boundary of the area where you want to deploy it.

One of the configurations observed in the use of urban gardens is the greenhouses. These are protected agricultural environments where the plastics are used as cladding materials for covering the greenhouse framework, used in the protection of crops, facing the climatic adversities. Agricultural greenhouses are used to create climatic environments suitable for plants, protecting them from poor environmental conditions such as frost, hail, and other weather. They are used for food production, cultivation of ornamental plants, flowers and medicinal plants [10].

Rosenzweig et al. [11] stated that the cultivation in protected environment brings with it numerous advantages such as: harvest in the periods between harvest, faster production cycle due to favorable environment conditions, increase in production, control of the environment promoting the development and production of plants, greater control of pests and diseases that may occur in the protected environment, better use of available resources, reduced risks and increased market competitiveness by the producer.

According to Wong [8], besides urban gardens, vertical farms have numerous advantages such as production of several crops throughout the year, zero loss of crops related to possible adverse weather conditions, reduction of transaction costs, production without pesticide use, herbicides and fertilizers, optimization of water resources, greater control of food security and social and esthetic gain in large urban centers.

2.1.2 Gardens in the school space

In the educational context, adopting other forms of production and consumption that is more sustainable is fundamental to resignify time, space and social relations of the students.

It is noteworthy that for school garden its size is not important, but the diversity of plants used matters. The main idea is to manage, in a balanced way, the soil and other natural resources through a work harmonized with nature and, as the garden will be located in the premises of a school, the construction and management can be used as an activity of enrichment of classroom learning [12].

The study proposes the elaboration for the construction of school gardens that meet children and adolescents for the experience of agricultural and environmental practices for plant and animal production. Vegetable gardens, vertical gardens in pet bottles and gardens in ceilings with the use of pallets are preferred in which vegetables and fruits adapted to the region, are to be cultivated. Proper management options are required, besides use of constructive materials of low cost. Community and rural producers in the region may supply the seedlings and may be involved in the production system.

There are enormous benefits of composting in the school community, and most important among them is the possibility of students to start administering and using the leftover food produced in their family environment. In addition are the inevitable learning process related to ethics, personal responsibility and environmental citizenship, giving them a specific action to help their local community and society as a whole. The landscape effect that the materials recyclable and plants can provide, has been shown in **Figure 1**.

With the short space, it was proposed to build units of vertical vegetable gardens using pet bottles that can be collected by students at their homes, vegetable gardens

Figure 1.
Proposal of an agropedagogical space of recyclable materials.

with tires and a central site that is already existing in space. For the cultivation of species that do not have good productive results receiving solar incidence, a recycle wood pergola of solid waste from civil construction—RSC, was designed where vines will be cultivated, for example, passion fruit.

Vegetable gardens serve as pedagogical space for teaching learning, almost all care is performed by the students, as seen in **Figures 2** and **3**.

Thus, the garden inserted in the school environment can be a living laboratory that enables the development of several pedagogical activities in environmental and food education, uniting theory and practice in a contextualized way, assisting in the teaching process, learning and promoting collective and cooperative work among the social agents involved [5].

It is worth noticing that this space will serve to perform different trainings that may be offered to the community, enabling knowledge about the breeding of birds, including the sanitary management, the preparation of the ration in the property, records and notes, the use of poultry manure for crop fertilization, production of vegetables and legumes and climatic studies, etc.

In the field study, one of the schools had the space in a covered area with roof in slab, and thus pallets were used for the construction of beds, in order to protect the floor of the covering and use a material that would possibly have its destination in the dumps.

Another proposal was on lajeed roofs, and again use of pallets was suggested (**Figure 4**). The pallets may be purchased in trade as recycled material, and used as the basis for transporting construction products, machinery, and other high-weight products.

In addition to providing a better nutrition to school students, it also ensured a greater awareness about the natural assets and valorous vision about the

Figure 2.
Planting with technical guidance.

Figure 3.
Vegetable bed with the use of tires.

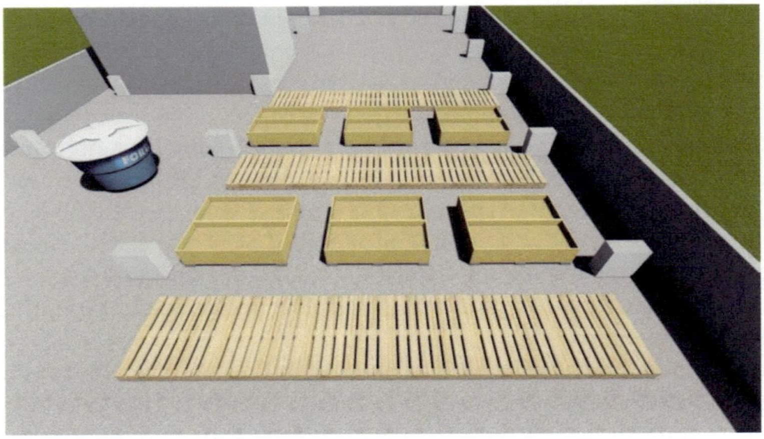

Figure 4.
Gardens on roofs utilizing pallets.

agricultural activities so present in their midst. The lack of encouragement to the young in the field generates the non-continuity of properties and the growing demographic, economic and cultural emptying of regions of predominance of family farming [3, 13]. According to Adeokun [1], all the efforts of the stakeholders for sustainable child development is valid, and this research continues to be a formidable way to rationalize alternatives and practices of agriculture in the school life.

Projects of this nature are of great relevance to transform some esthetic concepts such as the use of green spaces, actions geared towards environmental education, possibility of exploitation of reusable resources could be debated, used and transformed artistically in a vertical garden, which continues to be cared for by all students and school staff, as documented in the study of Oliveira et al. [6].

With pedagogical practices appropriate to the work, elaboration and development of the school garden in public schools, it is observed that there is also encouragement to the various forms of learning and understanding, enabling the acquisition of new knowledge, where all, through research and practice can exert a dynamic activity, which favors the teaching of science, enabling the encouragement of research and discussion of topics as a food environment, waste, cooperative work, behavior and make possible the development of the teaching-learning method, through practice, in addition to awakening social values such as participation, sense of responsibility, interpersonal relationship and awareness of the metastatic issues in the period in which we live.

3. Conclusions

With the study carried out, it was possible to construct different types of gardens for different spaces, bringing to the school spaces the plant production and the productive knowledge for children in urban areas, in the guidelines of (1) sustainability in the ecological, economic, social, cultural, political and ethical dimensions; (2) agricultural production bringing well-being and guaranteeing productivity; (3) construction with low cost and use of recyclable materials and adopting the method in educational spaces.

Author details

Adriana Maria dos Santos[1*], Mariana Paiva Baracuhy[1], Dermeval Araújo Furtado[1], Romulo Wilker Neri de Andrade[2], Jackson Rômulo de Sousa Leite[1] and Fabiana Terezinha Leal de Morais[1]

1 Federal University of Campina Grande, Campina Grande, Brazil

2 Universidade Federal da Paraíba, João Pessoa, Brazil

*Address all correspondence to: ttstadriana@gmail.com

References

[1] Adeokun OA. Research and agricultural extension support for sustainable development of Nigerian child: A general over view. Revista de Desenvolvimento Sustentável, Nigéria. 2017;**45**:11-18

[2] Brumer A. Gender relations in family – Farm agriculture and rural urban migration in Brazil. Latin American Perspectives. 2008;**35**(6):11-28

[3] Anjos FS, Caldas NV. Ser ou não ser agricultor? Eis a questão. Representações sociais sobre a profissão de agricultor entre jovens de comunidade rural do Sul do Brasil. Revista de Extensão e Estudos Rurais. 2015;**4**:23-28

[4] Guthman J, Agrarian D. The Paradox of Organic Farming in California, Available worldwide. California; 2014. p. 328

[5] Arruda RF, Marques MR, Reis JT. Implantation of school court using recyclable materials as an environmental education alternative. Interdisciplinary Scientific Journal. 2015;**4**(3):158-176

[6] Oliveira LHO, Abreu RF, Gobira MD, Assis GGD. Horta vertical: Um instrumento de educação ambiental na escola. Revista Eletrônica do Mestrado em Educação Ambiental, Rio Grande do Sul. 2014;Especial Impressa:4-14

[7] Dimoud A, Nikolopoulou M. Vegetation an urban environmet: Microclimatic analysis and benefits. Energy and Buildings. 2003;**35**(1):69-73

[8] Wong NH, Chen Y, Ong CL, Sia A. Investigation of thermal benefits of rooftop garden in the tropical environment. Building and Environment. 2005;**38**(2):261-270

[9] Arruda J. Agricultura urbana e periurbana em Campinas/SP: Análise do programa de hortas comunitárias como subsídio para políticas publicas [Dissertação]. Campinas: Mestrado em Planejamento e Desenvolvimento Rural Sustentável – Faculdade de Engenharia Agrícola, Universidade Estadual de Campinas; 2006

[10] Niachou A et al. Analysis of green roof thermal properties and investigation of its energy performance. Energy and Buildings. 2001;**33**(7):719-729

[11] Rosenzweig C, Gaffin S, Parshall L, editors. Green Roofs in the New York Metropolitan Region: Research Report. New York: Columbia University Center for Climate Systems Research and NASA Goddard Institute for Space Studies; 2006. pp. 5-15

[12] Santos ARS. Implantação Da Horta Escolar Em Uma Escola Pública Em Araras-SP. Medianeira: Universidade Tecnológica Federal do Paraná; 2009. p. 39

[13] Silvestro ML. Os Impasses Sociais da Sucessão Hereditária na Agricultura Familiar. 1st ed. Florianópolis: Editora EPAGRI; 2001. 96p

Section 2

B. Potential and Scope of Urban Horticulture

Chapter 7

Urban Horticulture in Sub-Saharan Africa

Ifeoluwapo Amao

Abstract

Horticultural crops refer to fruits, vegetables, spices, and ornamental and medicinal plants which are rich sources of vitamins, minerals, and phytochemicals. Rapid urbanization and migration of rural populace to the more industrialized city center has led to poverty, malnutrition, low and insecure incomes, ill-health and other livelihood problems. These problems are mostly seen among the people residing in urban areas who have migrated from rural areas. Urban horticulture ensures food and nutrition security, healthy environment and sustainable livelihoods, employment generation, among others. As such, this chapter carried out an empirical review of the state of urban horticulture in cities across sub-Sahara Africa. This is to enumerate ways whereby the benefits of urban horticulture can be specified in the region. It concluded that governments in the different countries need the political will to actualize identified benefits of urban horticulture. The chapter then recommends sensitization of the pertinent stakeholders in countries across sub-Saharan Africa on the benefits of urban horticulture. Such stakeholders include politicians, policy makers and urban households. This is in order to integrate the concept into urban land use planning while carefully considering sustainability of the environment.

Keywords: fruits and vegetables, sustainable city, food security, malnutrition, livelihood

1. Introduction

Low levels of poverty due to failure of growth in Gross Domestic Product (GDP) per capita has been observed in sub-Saharan Africa where the average real per capita income in year 2010 was 688 USD as compared to 1717 USD in other developing countries of the world [1]. Agriculture is a major contributor to GDP in Africa (up to 32%). Nevertheless, the sector is characterized by low productivity. It is important to increase agricultural productivity in Africa to ensure poverty reduction [1]. In addition to poverty, inhabitants of most cities in sub-Saharan Africa are experiencing micronutrient deficiencies. For instance, over 200 million and 1.6 billion suffer from Vitamin A and iron deficiencies respectively [2, 3]. In the same vein, it is expected that the populations in urban cities across sub-Sahara Africa will increase rapidly due to rural–urban migration and natural population increase [4]. A projection of 20.2% increase is expected by 2050 [5] which will place more burdens on the currently available food, fruits and vegetables in cities across the region. Other challenges posed by the rapid growth in these urban areas include food insecurity and unemployment [6]. Urban agriculture/horticulture is a veritable solution to these identified problems in the urban areas of sub-Sahara Africa.

There are various definitions of the term "urban"—an urban area can be determined based on a number of factors; for instance, population size, density, administrative function and other indicators such as infrastructure, facilities, employment [7].

Urbanization is the increase in number of people living in the cities caused by migration, commercialization, industrial growth and social factors (presence of educational facilities, better standard of living) [8].

Urban Agriculture (UA) can be defined as "an industry within (intra-urban) or on the fringe (peri-urban) of a town, city or metropolis which grows, raises, processes and distributes a diversity of food and nonfood products using largely human and material resources, products and services found in and around that urban area and in turn supplying human and material resources, products and services largely to that urban area" [9].

Fruits and vegetables form an essential part of horticultural crops rich in fiber, minerals and bioactive compounds. Consumption of fruits and vegetables is necessary to ensure healthy diets for balanced nutrition. They are consumed alongside staple foods and prevent diseases as a result of deficiencies across the population. Diets of most of urban dwellers are deficient in micronutrients such as vitamin A, iron, iodine, zinc [10].

The most widespread form of urban agriculture involves horticultural crops [11]. From their own viewpoint, Dimas et al. [12] defined urban agriculture as the production of fruits, vegetables and food within the urban environment for household consumption and sale. Considering the aforementioned, it can be inferred that urban horticulture is urban agriculture which has to do with production of horticultural crops—fruits, vegetables, spices, and ornamental and other medicinal crops. Thus, in this chapter, urban horticulture is urban farming with horticultural crops in focus.

2. Urbanization and urban agriculture/horticulture

Reasons for urbanization include employment opportunities (access to better paid jobs) and modernization. However, ills of urbanization have been identified to include crime, global warming, unhealthy environment, experienced in slums due to overcrowding and unemployment [8]. Population in some urban areas of West Africa has increased from 4% in 1920 to 45% in 2011. Urbanization has resulted in increased urban food demand challenging food production, rural–urban linkages, transport and traditional market chains [13]. A significant percentage (60–100%) of the perishable vegetables consumed within some African cities such as Ghana, Dakar, Bamako and Dar es Salaam, are produced through urban vegetable farming. This indicates a high contribution of urban vegetable farming to balanced diet of these urban dwellers [14].

Urban agriculture can be sustained with adequate development of urban areas. International policy makers have identified the role of urban and peri-urban fruit and vegetables production in enhancing vitamin and micronutrient supply for households in the urban areas especially the poor ones [15].

Urban agriculture can increase employment and income which brings about the ability to purchase food and increase the diet of households thereby ensuring food security. For example, in Ghana during the post-independent economic crises, the government supported urban agriculture as a means of meeting the population's food demand by launching the "Operation Feed Yourself" programme in which the urban population were encouraged to practice aquaculture, plant everywhere and anywhere in the cities [16]. There were 800–1000 farmers in Accra (Ashiaman-Tema area) with 60% of them producing exotic vegetables while the remaining

40% produced indigenous vegetables. They produced exotic crops such as lettuce, cabbage, cauliflower, spring onions and indigenous crops such as okra, *Corchorus* spp., aubergine, hot pepper and tomato. These crops were grown within the city on plot size ranging between 0.01–0.02 ha per farmer [13].

Moreover, urban agriculture is an important food security strategy for urban households [9, 17, 18] since it improves access to a ready fresh crops [19] rich in essential micronutrients in poor household diets [20, 21].

Urban horticulture provides highly nutritious and healthy plant-based foods; it also serves as a means of securing the livelihood of urban population. For instance, more than 70% urban rowers in Tamale, Ghana are involved in vegetable production for home consumption and the market [22].

It is important to include urban horticulture in urban land use planning and policy making because if well managed, it will serve as an important tool for poverty reduction, environmental management and economic development in most developing countries. Several stakeholders are required to come together to achieve the benefits of urban horticulture including politicians, legislators, urban planners, land owners, entrepreneurs, producers and urban dwellers. These stakeholders act at the local, national and international levels to transform the concept to operational standards and actions which will enable it to contribute to food security, food safety and livelihoods.

Participating in urban horticulture improves climatic factors, e.g., biodiversity, air quality, water management. However, cultivation close to major roads and railways as well as abandoned sites should be avoided as they pose threats of health hazards to the consumers through contamination of the produce [23, 24]. Thus, the health implication of potential hazards obtained from producing fruits and vegetables in urban areas should not be neglected.

Moreover, if urban horticulture is given more technical and institutional support, the sector may assist expanding African cities in achieving zero hunger [25].

Integrating urban agriculture into land use planning gives rise to urban greening, open green spaces, urban habitat diversity as well as reduction in noise and pollution. Economically, it will lead to food security and community revitalization through participation in community gardening. Income generation and employment creation can also be achieved through urban agriculture. In order to have a sustainable city, basic amenities such as water and waste have to be produced and managed in line with the principles of sustainable management, that is, economic, environmentally friendly and equitable [26].

3. Empirical reviews on urban agriculture/horticulture

An exploratory cross-sectional survey was carried out in Eldoret, Kenya to examine the effect of socio-economic characteristics of low-income horticultural food producers and sellers on their livelihoods and household food security. The results showed that male respondents were more involved in production while females were more involved in selling. Sellers were more food secure than producers while producers felt safer to make available quality food for their households when they experience tough food situations. Producers and sellers affect each other's livelihoods while the success of one group leads to that of the other. It is imperative to increase farming in the city in order to meet the employment and food needs of the urban population especially the poor [15].

Mkwambisi et al. [27] in their study, empirically examined the role of urban agriculture in urban households with focus on two main cities in Malawi- Lilongwe and Blantyre. The study observed that on the average, households surveyed could support themselves on the food they produced on the urban agriculture plots which were either within their living houses or on plots around the urban area. The study also revealed that more educated, wealthier and male-headed households had significantly more harvest than poorer, less educated and female-headed households. However, the female-headed households obtained more income from urban agriculture than their male counterparts. However, they are more likely to sell horticultural crops than the female-headed households. Vegetable production was the most lucrative sector for the sampled urban households; other sectors involved were arable maize, livestock and poultry production. The study further revealed that urban agriculture is the second most important source of income for the sampled households. Households produced horticultural crops due to their low demands of expensive fertilizer and short production cycle. In addition, the study observed the need for promoting policy that could support the production of vegetables in urban farms, as well as encourage contract farming so as to link urban farmers to high value markets to provide them with high income and also a means of employment and wealth creation. It concluded that urban agriculture should be recognized on the political agenda in order to realize its immense benefits. This is because those involved in urban agriculture were important stakeholders in a bid to tackling the issue of food insecurity and poverty faced in most sub-Saharan African cities. Urban sustainability could be achieved if municipal solid waste and waste water are used to produce food and livestock.

In eleven Southern African cities, Frayne et al. [28] performed a study to confirm the potential benefits of urban agriculture in urban development and poverty alleviation under the present practice and regulations. Secondary data collected by the African Food Security Urban Network (AFSUN) between year 2008 and 2009 from 6453 households residing in the selected Southern African cities was used for the study. The results revealed that no significant differences existed between households who were engaged in urban agriculture as a food source and those who did not. However, there were exceptions in Maseru (Lesotho) where households involved in urban agriculture as a food source had significantly better access to food and dietary diversity than other households in the city. The same trend was observed in Lusaka (Zambia), Cape town and Johannesburg (South Africa). Also, the practice of urban agriculture in the study areas is not an effective strategy for food security despite variations observed. Urban agriculture is correlated with education, wealth and landholding of the household head (as reported in [27]). Furthermore, findings revealed that urban agriculture plays limited roles in poverty alleviation. The rate of household engagement is determined by political and historical circumstances. No significant relationship was observed with urban agriculture and food security in most cities. It was concluded that significant investment and support in terms of inputs, extension services, credit access, production and marketing infrastructure are all required to be able to realize the potential benefits of urban agriculture for food security and poverty alleviation.

Moustier [29] examined the role of urban horticulture in contributing to the supply of vegetables in African and Asian cities. Findings revealed that to provide perishable food items such as fruits (plantain/banana) and fresh perishable vegetables in the study locations, urban agriculture is important. Some examples of fresh perishable vegetables produced in African and Asian cities are amaranth, cabbage, lettuce, etc. The study also showed that producing close to the point of consumption has a dual advantage of reducing physical transport cost as well as information and transaction cost related to marketing. This is beneficial as it guarantees food safety. In addition, the study noted that for public support into urban agriculture, the government should include financial support, integrate urban agriculture/horticulture into urban planning, encourage innovative marketing and quality labeling as well as ensure research and extension to improve profit and ensure sustainability of intensive commercial vegetable and animal systems.

The contribution of urban agriculture to socioeconomic development of urban dwellers was assessed in three urban centers of Nasarawa state [30]. The socioeconomic characteristics of urban farmers sampled showed that 60% were aged 41–60 years, 55.56% were female, and 90% were married. Moreover, most of them (77.78%) produced vegetables, maize (66.67%), ornamental crops (61.11%). A greater proportion of the urban farmers derived additional income from their farming activities while 55.56% opined provision of household feeding as a benefit of urban farming. Constraints faced by this group of farmers were poor extension service, low capital, high cost of labor, inadequate input supply and land, theft of produce and products as well as encroachment of farms. The study then concluded that urban agriculture should be considered in urban land use planning since it is a source of urban income, employment and food systems. Also, urban agriculture should be integrated into national agricultural research in order to achieve intensive and sustainable cropping systems.

A study carried out by Ibok et al. [31] examined the productivity of urban food crop farming households and its effect on their food security status. The study collected data from three urban centers in Cross River state, Nigeria. There were more food insecure households (53.5%) than food secure ones (46.5%) involved in the study. The productivity of urban farming households positively and significantly affected the food security status of households.

Adedayo and Tunde [32] studied the motivation for women involvement in urban agriculture in Kwara state Nigeria. Women were attracted to this activity considering food security, access to land and income supplement. As such, income realized from urban farming could then be used to meet other basic needs. The study noted that owing to its potentials, urban agriculture should be encouraged in both small and big towns in Nigeria and other developing countries.

Obuobie and Sarpong [33] conducted an informal study on irrigated vegetable production in urban and peri-urban areas of Cape coast and Takoradi, Ghana. Exotic vegetables found in the area included spring onion, cabbage, carrot, lettuce and cucumber. Insignificant level of irrigated vegetable production was observed in Cape coast due to the saline nature of soil in the area, scarcity of fresh water and unsuitable topography. On the other hand, considerable quantities of vegetables were produced in different production sites across Takoradi. The study found that about 25 farmers were involved in the activity, all of whom were males and aged 28–45 years with about 5 years experience in production of vegetables such as cabbage, carrot, spring onions, cauliflower, lettuce and green pepper. Other vegetables they produced were tomatoes, okra, long bean and black bean. These crops were being produced utilizing mostly family labor on an average farm size of 0.12 hectares. Farmers experienced low pricing for their produce as a result of glut from the sales of similar products that were brought in from outside the urban and peri-urban area. There was also lack of storage system in the peak season which led to perishability of the produce. Despite this, some of the farmers attested that they earned a gross margin of about 2 million Ghana cedis. The study concluded that if the problems mitigating irrigated vegetable production in Cape coast is addressed, the youth will be encouraged to earn a living through this venture, i.e., irrigated vegetable production.

Furthermore, Tornyie [26] examined critical approaches needed to incorporate urban agriculture into urban planning and management in two urban cities—Accra and Kumasi in Ghana. The study revealed that the farmers involved in urban agriculture produced crops such as spring onion, lettuce, cabbage, green pepper, okra, cauliflower. The activity was male-dominated in both study locations; farmers were literate and some employed other people to assist in their production activities. The main reasons they were engaged in the activity was availability of water and profit being realized. Farmers experienced constraints such as inadequate access to credit as well as safe and cheap irrigation facilities, limited access to land, threat from pest and diseases, and marketing of produce. Moreover, findings showed that no comprehensive plan existed for urban agriculture; the issue was partially mentioned in the bye law of Accra Metro Assembly of 1995. This supported backyard farming and was promulgated with the major aim of maintaining sanitary conditions in the metro assembly and not to promote urban agriculture. The study recommended that good physical planning should not regard agricultural lands as residential lands or those for commercial uses. Also, integrating urban agriculture into city development would require including urban and peri-urban agriculture in zoning plans, construction of urban territorial way, among others. In addition, policy makers should ensure that the needs and benefits of urban agriculture (land tenure, drainage and water availability) should be considered in physical planning.

In addition, the socio-spatial dynamics of household food and nutrition security was assessed by examining vegetable production, consumption and its contribution to diets of households [34]. This was carried out to assess the role of urban and peri-urban agriculture in Tamale, Ghana. The results revealed that okra, pepper and roselle were the most commonly produced vegetables in the study area cultivated mainly for household consumption. A greater proportion of the households produced staple crops compared to vegetables. The results also showed a low consumption of dark green leafy vegetables and limited diversity of vegetables, especially vitamin A rich vegetables and tubers. Urban households had highest dietary diversity and dark green leafy vegetable consumption. This finding could be due to their accessibility to fruit and vegetable markets as well as other food suppliers. However, limited diversity of vegetables consumed was noted in the rural areas.

Lastly, Mugalavai et al. [15] concluded that in Eldoret, Kenya, increasing farming in cities was important to meet employment and food needs, especially for the poor urban population. From the findings of Mkwambisi et al. [27], they opined that urban agriculture should be recognized on the political agenda. This was because it was the second most important source of income for households in Blantyre and Lilongwe, Malawi. The need for investment and support for urban agriculture in terms of extension services, credit access, and so on was observed [28]. The study also observed the need to integrate urban agriculture into urban land use planning [26, 30].

4. Conclusions

The identified benefits of urban horticulture include food security, employment and income generation for producing households as well as individuals employed to work on the urban farms. It also reduces the cost of transportation and easy access of healthy food to the teeming urban population.

In order to realize the benefits of urban horticulture in the sub-Saharan African region, it is imperative that governments in the different countries have the political will to actualize the identified benefits. This can be achieved by sensitizing them on the benefits of urban horticulture and integrating this concept into urban land use planning while ensuring environmental sustainability.

Conflict of interest

The author declares no conflict of interest.

DOI: http://dx.doi.org/10.5772/intechopen.90722

Author details

Ifeoluwapo Amao
National Horticultural Research Institute, Ibadan, Nigeria

*Address all correspondence to: ifeluv@yahoo.com

References

[1] Chauvin ND, Mulangu F, Porto G. Food Production and Consumption Trends in Sub-Saharan Africa: Prospects for the Transformation of the Agricultural Sector United Nations Development Programme (UNDP) Working Paper WP 2012-0112012. p. 76

[2] WHO. Worldwide Prevalence of Anaemia 1993-2005: WHO Global Database on Anaemia. Geneva, Switzerland: WHO; 2008

[3] WHO. Global Prevalence of Vitamin a Deficiency in Populations at Risk 1995-2005: WHO Global Database on Vitamin a Deficiency. Geneva, Switzerland: World Health Organization (WHO); 2009

[4] Prain G, Lee-Smith D, Karanja N, editors. African Urban Harvest: Agriculture in the Cities of Cameroon, Kenya and Uganda. New York, Ottawa, Lima, Peru: Springer, International Development Research Centre, International Potato Center; 2010. p. 3

[5] UN-HABITAT. The State of African Cities 2014: Re-Imagining Sustainable Urban Transitions. Nairobi, Kenya: United Nations Human Settlements Programme; 2014. p. 23

[6] UNFPA. The State of the World Population. Unleashing the Potential of Urban Growth. New York: United Nations Population Fund; 2007. pp. 1-108

[7] McGranahan G, Satterthwaite D. Urbanisation Concepts and Trends. International Institute for Environment and Development (IIED) Working Paper. London: IIED; 2014. Available from: http://pubs.iied.org/10709IIED. ISBN 978-1-784 31-0 6 3-9

[8] Pawan. Urbanization and its causes and effects: A review. International Journal of Research and Scientific Innovation. 2016;**III**(IX):110-112

[9] Mougeot LJA. Urban agriculture: Definition, presence, potentials and risks. In: Bakker N, Dubbeling M, Guendel S, Koschella US, de Zeeuw H, editors. Growing Cities, Growing Food, Urban Agriculture on the Policy Agenda. Germany: DSE; 2000. pp. 1-42

[10] Tenkouano A. The nutritional and economic potential of vegetables. In: The Worldwatch Institute, editor. State of the World 2011: Innovations that Nourish the Planet. New York, London: W.W. Norton and Company; 2011. pp. 27-37

[11] Ruggeri G, Mazzocchi C, Corsi S. Urban gardeners' motivations in a Metropolitan City: The case of Milan. Sustainability. 2016;**8**:1099. DOI: 10.3390/su8111099

[12] Dimas SJ, Ogunmokua AA, Nantanga T. The status of urban and peri-urban agriculture. A Survey Report Prepared for Integrated Support to Sustainable Development and Food Security Programme in Namibia; 2002

[13] Drechsel P, Keraita B, editors. Irrigated Urban Vegetable Production in Ghana: Characteristics, Benefits and Risk Mitigation. 2nd ed. Colombo, Sri Lanka: International Water Management Institute (IWMI); 2014. p. 247. DOI: 10.5337/2014.219

[14] Zezza A, Tasciotti L. Does urban agriculture enhance dietary diversity? Empirical Evidence from a Sample of Developing Countries (draft). 2008. Available from: ftp://ftp.fao.org/docrep/fao/011/aj304e/aj304e.pdf

[15] Mugalavai VK, Omutimba D, Harriet N. Urban agriculture livelihoods and household food security: A case of

Eldoret, Kenya. Journal of Agricultural and Biological Science. 2013;**8**(1):90-96

[16] Asomani-Boateng R. Urban cultivation in Accra: An examination of the nature, practices, problems, potentials and urban planning implications. Habitat International. 2002;**26**:591-607

[17] Klemesu MA, Maxwell D. Accra: Urban agriculture as an asset strategy supplementing income and diets. In: Barker N, Dubbelling M, Gindel S, Sabel-Koschella U, de Zeeuw H, editors. Growing Cities, Growing Food-Urban Agriculture on the Policy Agenda. Eurasburg, Germany: DSE; 2000. pp. 183-208

[18] Nugent R. The impact of urban agriculture on the household and local economies. In: Barker N, Dubbelling M, Gindel S, Sabel-Koschella U, de Zeeuw H, editors. Growing Cities, Growing Food-Urban Agriculture on the Policy Agenda. Eurasburg, Germany: DSE; 2000. pp. 67-97

[19] Mwangi A, Foeken D. Urban agriculture, food security and nutrition in low income areas of the city of Nairobi, Kenya. African Urban Quarterly. 1996;**11**(2-3):170-179

[20] FAO. Urban Agriculture and Peri-Urban Agriculture: A Briefing Guide for Successful Implementation of Urban Agriculture and per-Urban Agriculture in Developing Countries and Countries of Transition. Rome: FAO; 2001

[21] Maxwell DG. The Importance of Urban Agriculture to Food and Nutrition. Annotated Biography by ETC-RUAF. Wageningen, Netherlands: CTA Publishers; 2001

[22] Abubakari AH, Mahunu G. The role of urban horticulture in urban development – A case study of the tamale Metropolis. Ghana Journal of Horticulture. 2007;**6**:129-133

[23] Khan S, Caoa Q, Zhenga YM, Huanga YZ, Zhu YG. Health risks of heavy metals in contaminated soils and food crops irrigated with wastewater in Beijing, China. Environmental Pollution. 2008;**15**:686-692

[24] Nabulo G, Young SD, Black CR. Assessing risk to human health from tropical leafy vegetables grown on contaminated urban soils. Science of the Total Environment. 2010;**408**(22):5338-5351

[25] FAO. Growing greener cities in Africa. In: First status report on urban and peri-urban horticulture in Africa. Rome: Food and Agriculture Organization of the United Nations; 2012. p. 116

[26] Tornyie EK. The Integration of Urban agriculture into sustainable city development, a case study of Accra and Kumasi cities [thesis]. School of Graduate Studies, Kwame Nkrumah University of Science and Technology, Kumasi, College of Architecture and Planning, Department of Planning; 2011. p. 113

[27] Mkwambisi DD, Fraser EG, Dougill AJ. Urban agriculture and poverty reduction: Evaluating how food production in cities contributes to food security, employment and income in Malawi. Journal of International Development. 2011;**23**:181-203. DOI: 10.1002/jid.1657

[28] Frayne B, McCordic C, Shilomboleni H. Growing out of poverty: Does urban agriculture contribute to household food security in southern African cities? Urban Forum. 2014;**25**:177-189. DOI: 10.1007/s12132-014-9219-3

[29] Moustier P. Urban horticulture in Africa and Asia: An efficient corner food supplier. Acta Horticulturae. 2007;**762**:145-148

[30] Salau ES, Attah AJ. A socio-economic analysis of urban agriculture in Nasarawa state, Nigeria. Production Agriculture and Technology Journal. 2012;**8**(1):17-29

[31] Ibok OW, Idiong IC, Bassey NE, Udoh ES. Food security and productivity of urban food crop farming households in southern Nigeria. Agricultural Sciences. 2014;**2**(3):1-12. DOI: 10.12735/as.v2i3p01

[32] Adedayo A, Tunde AM. Motivations for women involvement in urban agriculture in Nigeria. Journal of Agriculture and Rural Development. 2012;**2**(3):337-343

[33] Obuobie E, Sarpong E. General overview of Urban and Peri-urban Agriculture in the Cape Coast and Takoradi Municipalities, Urban and Peri-urban Agriculture (UPA) Studies. West Africa, Ghana: International Water Management Institute (IWMI); 2005. pp. 1-8

[34] Chagomoka T, Drescher A, Glaser R, Marschner B, Schlesinger J, Nyandor G. Vegetable production, consumption and its contribution to diets along the urban–rural continuum in northern Ghana. African Journal of Food, Agriculture, Nutrition and Development. 2015;**15**(4):10352-10367

Chapter 8

Soil Quality Problems Associated with Horticulture in the Southern Urban and Peri-Urban Area of Buenos Aires, Argentina

Paladino Ileana, Sokolowski Ana Clara,
Prack Mc Cormick Barbara, José Enrique Wolski
and Rodríguez Hernán y Mauro Navas

Abstract

Horticulture is the main productive activity of south Buenos Aires city peri-urban sector. This activity is carried out with intensive land use, based on the high use of inputs, which has generated important pollution and soil degradation problems. Soil degradation processes have their origin in the poor quality irrigation water (sodium bicarbonate) and in the indiscriminate use of fertilizers and organic fertilizers, without considering the requirements of the crop and soil analysis. The results of a large number of surveys in the area, specified in the following chapter, showed salinization, pH increase, structure quality loss, organic matter decrease and phosphorus hyperfertilization. On the other hand, urban gardens are increasingly common, that is, the production of vegetables for own consumption within the urban framework. In this case, the problems are related to the type of soils where it occurs, and they are in general highly modified lands that almost completely lost their natural characteristics and are usually not favorable for plant growth. The results from the cases studied in La Plata city showed that urban soils have low organic carbon content, high bulk density and high pH. In these soils, the horticultural production with agroecological base managed an increase in the organic carbon content and a decrease in the apparent density.

Keywords: soil quality, anthropic soils, agroecology, overfertilization, soil degradation

1. Introduction: characteristics of the horticultural sector

Argentina has a continental area of 2.8 million km^2 from which about 34 million hectares are destined to agricultural crops production. Vegetables and legumes production occupies only 1.5% of that total; however, it represents 10% of the gross agricultural product [1]. Horticulture is characterized by its high degree of intensity in the use of land, labor, capital and technology, so it has social importance and generates a large number of jobs [2]. On the other hand, because this activity is carried out in every province of Argentina, it has importance from a geopolitical and strategic point of view, being part of the "regional economies" [1].

The wide distribution of horticulture in Argentina is due in part to the diversity of climates that it possesses, however the commercial production that supplies the main urban centers with fresh vegetables is located in the peripheral area of those urban centers (peri-urban). The peri-urban is represented by an area of small farms and orchards in the surrounding of large cities, which production is specialized in leafy vegetables and seasonal vegetables. Therefore this area is commonly referred to as the "green belt" [3]. The green belt of Buenos Aires city has more than 5.510 km^2 and includes a population of more than 4.5 million habitants. This area is represented by the districts of La Plata, Florencio Varela, Berazategui, Almirante Brown, Esteban Echeverría, La Matanza, Merlo, Moreno, Cañuelas, General Rodríguez, Luján, Marcos Paz, Pilar and Escobar. The three most important districts in terms of horticulture are in the southern peri-urban of Buenos Aires city (La Plata, Berazategui and Florencio Varela) (**Figure 1**). These districts represent 82% of the total horticultural farms and 81% of the surface under cultivation [5]. This is one of the most important areas for fresh food production (refer to leafy vegetables or flowers, fruits and stems); therefore, horticulture is very important in the local economy.

Horticulture in the southern green belt of Buenos Aires city is characterized by having an intensive and highly diversified production scheme in terms of production systems and diversity of species cultivated. The largest area in hectares is under open-field cultivation systems, however most producers have a combination of both open-field and greenhouse cultivation systems. The total area of the southern green belt under vegetables production is 5332.8 ha, with 30% under greenhouses cultivation systems and the rest in open-field cultivation systems [4]. Although there is a marked heterogeneity of producers, small and medium-sized enterprises, mainly from family farming, stands out. However, there are also a small number of business-type productions with mainly hired labor. Approximately 70% of producers rent small areas of land (smaller than 5 ha) and can have 1–1.5 ha under greenhouses [6]. Only half of the producers have technical advice mostly from

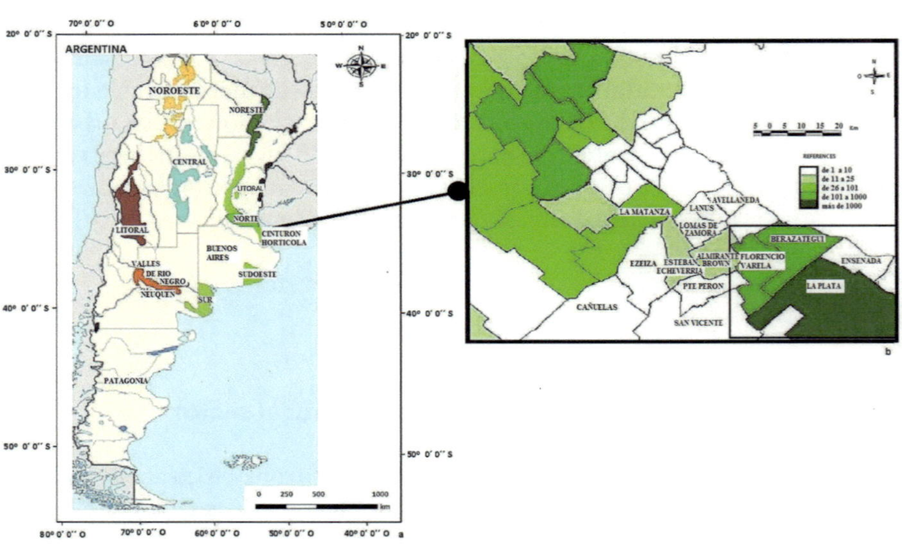

Figure 1.
(a) Argentinian horticultural regions. (b) Buenos Aires metropolitan districts that integrate de Green belt. References indicate the number of farmers per district. Source: [1] and Censo Hortiflorícola de Buenos Aires [4] Ministerio de Asuntos Agrarios—PBA. The boxed area indicated the sampling area (Florencio Varela and La Plata).

private sources, and to a lesser extent from public sources [1]. The main vegetables grown in greenhouses are: (in order of importance) *Solanum lycopersicum* (tomato), *Spinacia oleracea L.* (spinach), *Lactuca sativa* (lettuce), *Capsicum annuum* (pepper), among others of minor importance. The main vegetables grown in the open-field are: *Lactuca sativa* (lettuce), *Beta vulgaris var. cicla* (chard), *Spinacia oleracea L.* (spinach), *Solanum lycopersicum* (tomato), *Capsicum annuum* (pepper), *Cucurbita pepo* (trunk zucchini), *Brassica oleracea var. italica* (broccoli), *Brassica oleracea var. capitata* (cabbage), among others [7].

2. Climate and soil in the southern Green belt of Buenos Aires

The climate is mild without a dry season, with hot summers and mild winters. The average annual temperature is 16°C and the frost-free period is of 220 days (from October 20 to May 10). With the expansion of greenhouses, the influence of climatic conditions has diminished, however, the extreme temperature, high air relative humidity and the excess or deficit of light represent climatic limits for this production system. The total rainfall in the area is between 900 and 1000 mm a year and is distributed more or less uniformly in the four seasons. The water used for irrigation is extracted from between 55 and 60 m deep in each farm [1]. Generally the irrigation system is drip irrigation (20 cm between drippers, with two irrigation hoses per spine) and fertilization is performed by fertirrigation [8].

With respect to geomorphology, the southern green belt is characterized by gentle undulations of long and not very steep slopes. The area that encompasses the southern urban and peri-urban area is crossed by streams that run through it and drain into two basins: the Río de la Plata and the Samborombón and the Salado rivers. Between these basins there is a quite elevated terrain that separates their drainage.

The types of soil that predominate vary with the origin of their primary materials. In the higher areas, lithologically, the original materials belong mostly to the

Soils characteristic	La Plata	Florencio Varela
BD (g cm^{-3})	0.96	1.08
AS (MWD)	2.29	2.57
Soil texture	Clay loam-silty clay loam-silty loam	Silty clay loam-silty loam
pH	7.15	6.96
CE (ds m^{-1})	0.21	0.16
TOC (%)	2.86	3.21
POC (%)	0.24	0.73
MOC (%)	2.62	2.48
TN (%)	0.38	0.33
EP (mg kg^{-1})	81.40	80.26

BD: bulk density by cylinder method [15]; AS (MWD): aggregate stability (mean weight diameter) [16]; Soil texture: clay, sand and silt percentage (USDA triangle); pH and EC: electrical conductivity in water 1.2.5 by potentiometry [17]; TOC: total oxidative carbon by Walkley and Black [18]; MOC: oxidative carbon associated with the mineral fraction; POC: particulate oxidative carbon [19]; TN: total nitrogen by Kjeldahl micro-method [20]; and EP: extractable phosphorus by Bray and Kurtz 1 [20].

Table 1.
Peri-urban soils characteristic in non-cultivated sites.

"pampean sediment" consisting of sand-clay silt with limestone (loess). However, the limestone is not present in the upper horizons due to washing. Properties as pH, organic matter and soluble salts make these soils suitable for agriculture. In addition, due to their topographic condition they are not affected by floods [9]. Most agricultural soils in the southern green belt belong to the Molisol and Vertisol orders [10] (Chernozems and Vertisols) [11]. They have a strong profile development with dark A horizons, generally thick and well provided with organic matter followed by B horizons enriched in eluviated clay accompanied, especially in Vertisols, by evidence of expansion and contraction of materials. These are soils with high cation exchange capacity given by both organic matter and clay content. From the physical point of view, high levels of clay and silt, makes them susceptible to changes in their physical properties. In some cases there is low permeability and high plasticity in B horizons.

Low permeability is the main physical limitation for soils management in this area being more pronounced in the district of La Plata, where Vertisols with high clay content in surface (between 32 and 40%) and depth (50–60%), prevail. The only chemical limitation is low P content (less than 10 mg kg^{-1}) [12]. However, [9] reported 20.7 mg kg^{-1} in soils from natural fields. Moreover, [13, 14] studying the surroundings of La Plata reported levels of extractable P greater than 20 mg kg^{-1} in soils that were not under agriculture for at least the past 20 years. **Table 1** describes some physical and chemical properties of soils around the houses of horticultural farmers from the southern green belt of Buenos Aires city.

3. Traditional horticulture in the southern Green belt of Buenos Aires

Horticulture depends on the soil; therefore the conservation of this resource is essential to ensure the development of the sector. The intensive use of the soil, based on the overuse of inputs for many years (fertilizers, pesticides, disinfectants), added to irrigation with low quality water (sodium bicarbonate) [21], has generated negative impacts associated with soil pollution and degradation. The alteration of soil physical, chemical and biological properties resulted in a fragile productive system with a high risk of environmental impact [22, 23].

Soil physical condition establishes its sustainability, root penetration, air circulation, water storage capacity, drainage, nutrient retention, among other factors. The texture (proportion clay, silt and sand) determine the amount and availability of water, nutrients, aeration and drainage. Soils with fine texture (clay more than 40% or silt more than 60%) have a high water and nutrient retention capacity. However, it must be handled with caution because they are easily compacted. Likewise, crops sensitive to soil pathogens are more susceptible in heavy textured soils. Fertirrigation has counteracted these disadvantages. Soil compaction reduces porosity and increases its bulk density. This limits the space for the storage or movement of air and water within the soil. It is the cause of a physical restriction for the germination and radical growth of crops. Low values of organic carbon, high humidity and fine textures are more susceptible. The appropriate values for the growth of any plant with these textures must not exceed 1.1 g cm^{-3}. Furthermore, a good structural stability helps keep particles together against different destabilizing forces that can act on a soil. Soil chemical condition like pH influences nutrient availability and soil microbial activity. In acid soils, few nutrients are available to be taken by the roots. Suitable values for horticultural crops are between 6 and 7.5 depending on the crop. The electrical conductivity determines the concentration of soluble salts present in the soil solution and can affect germination, plant growth or water absorption. Each crop has a different capacity to support soil salinity without experiencing detrimental effects on its development and production (tolerance). The organic carbon brings life to the soil, is a nutrient reserve, improves its physical conditions, improves soil structure and its porosity, increasing the aeration and water circulation that favors the development of the plant, regulates microbiological activities; privileges infiltration, decreasing soil erosion, improving soil water balance, tends to reduce evaporation and is a water reservoir. Well-supplied soils have carbon values between 4.5 and 6.5% [24].

The main degradations observed are edaphic salinization and alkalinisation, which are associated with waterlogging, development of pests and nutritional imbalances [13, 25–27]. These observations are partially coincident with those reported for other regions of the country [28] and the world [29–31].

Soils management and the use of groundwater for irrigation, do not often consider local properties, but are mainly based on information obtained in other parts of the world with characteristics very different from those described in the area of study. In general, there is a lack of knowledge and awareness of the processes involved in soil, air and water pollution. Producers from the southern green belt do not perform soil analysis prior to sowing or fertilizing, the quality of the irrigation water is unknown and soil management is performed based on the particular experience of each producer [1]. In many cases a "fertilization recipe" is applied without considering the physicochemical characteristics of the soil of each farm and without considering the nutrient demand of the crop [32].

Salts and sodium accumulation is exacerbated in soils with higher clay content since in these cases drainage is limited and therefore salts are not eluviated. Comparing 17 soil samples, obtained randomly from not cultivated sites (NC) placed within the horticultural farms (e.g., areas adjacent to the houses), in two districts from the southern green belt, it was found that soils from the district of La Plata (with higher clay content on the surface horizon) had higher electrical conductivity compared with soils from Florencio Varela (**Figure 2**). The crops present in the most of the sampled soils were chard, lettuce, onion, crucifixes and zukini.

Likewise, salinization is also exacerbated in greenhouse cultivation systems with respect to open-field cultivation systems. In greenhouses, the excessive application of fertilizers, the use of low quality water with high content of sodium bicarbonate for irrigation, evapotranspiration that favors the accumulation of salts on the surface and the impossibility of washing with rainwater, further complicates the situation. Formerly, it was a common practice to remove the ceilings of the Greenhouses, made of glass, and allow the soil to be "washed" by the rain. Nowadays, this practice is less frequent, mainly because the cover of the greenhouses has been changed from

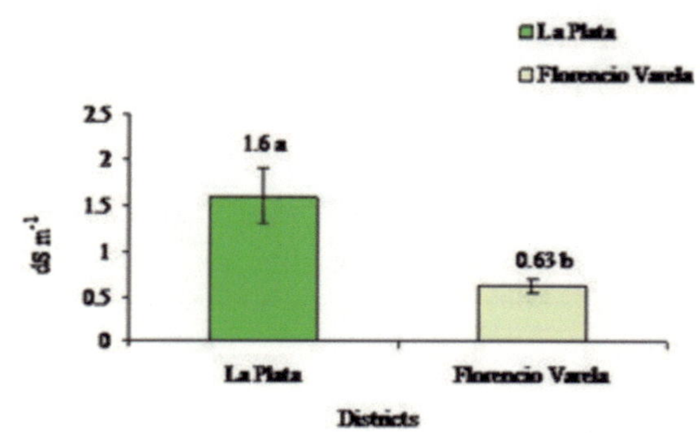

Figure 2.
Soil electrical conductivity in not cultivated soils per district. EC: electrical conductivity in water 1:2.5 by potentiometry [17]. Vertical bars indicate the standard deviation. Different letters show significant differences between treatments according to Tukey ($p < 0.05$) [33].

glass to plastic [13]. Thirteen soil samples were randomly obtained from horticultural farms from Florencio Varela district and it was observed that the electrical conductivity was higher in greenhouses than in open-field cultivation systems (**Figure 3**).

Although the electrical conductivity levels found in this study are still adequate for the growth of most horticultural species, some of them are at the limit. Vázquez and Terminiello (2008) found that saline concentrations greater than 1.2 ds m^{-1} affect the development and, consequently, the yield of onion, with a production reduction of 16% for every additional ds m^{-1}. Moreover, an adequate electrical conductivity should not exceed 1.5 ds m^{-1} for pepper [34] and 1.5 ds m^{-1} for tomato [35, 36]. [8] in a study carried out in greenhouses from La Plata district, found a high number of sites with high salinity levels with extremes that reached 10.6 ds m^{-1}. Under salinity conditions, there is a nutritional imbalance, due to different factors such as nutrient availability and competitive absorption.

Alkalization also represents an important problem in horticultural production. Cultivable soils from the southern peri-urban are naturally neutral or slightly alkaline but their alkalinity increases with horticultural use [9, 37]. Soils from the southern green belt were studied and it was observed that cultivated soils have higher pH, in both open-field and greenhouse cultivation systems, in comparison with non-cultivated soils.

It is important to consider that under high pH levels the availability of certain nutrients will be diminished, affecting crop development [38]. pH levels found in this study exceed the tolerance thresholds for many of the horticultural species grown in the southern green belt [39]. In this regard, [40] established that high pH levels (greater than 8) decrease the absorption of P and Ca^{+2} by crops, and on the contrary increase that of Mg^{+2}, because the absorption of its competitive elements diminishes. Increases in pH and electrical conductivity in cultivated soils compared to uncultivated ones are mainly linked to the quality of the water used for irrigation. **Table 2** shows results from irrigation water analysis, sometimes also used for human consumption. The samples were taken from horticultural farms from the southern green belt of Buenos Aires. The results show that in addition to the low quality in terms of physicochemical characteristics, water samples have also microbiological contamination with faecal bacteria. This usually happens due to the high number of improperly drilled holes that cause leaks from nearby blind wells. Inadequate water management brings associated problems such as lower

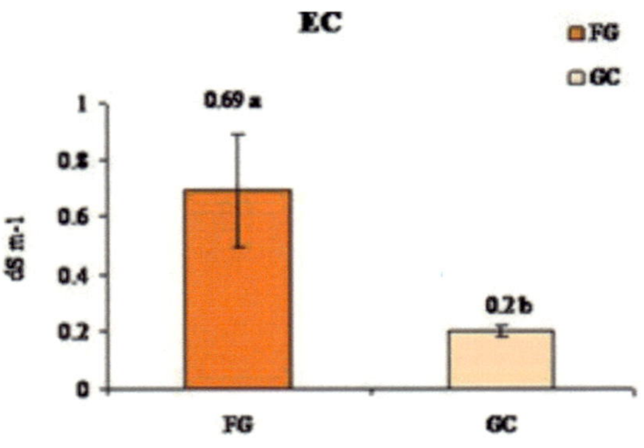

Figure 3.
Electrical conductivity values per system. EC: electrical conductivity in water 1:2.5 by potentiometry [17]; FC: open-fields cultivation systems; GC: greenhouses cultivation systems. Vertical bars indicate the standard deviation. Different letters show significant differences between treatments according to Tukey ($p < 0.05$). [33].

Well depth	Count		Investigation		Nitrites	Nitrates	pH	EC
m	MAB u.f.c. ml^{-1}	TC NMP 100 ml^{-1}	*E coli* in 100 ml	PA in 100 ml	ppm	ppm		µs cm^{-1}
60	40	5.1	+	0	0.01	50	8.4	810
50	75	16.1	+	0	0.01	50	7.8	870
75	185	5.1	+	0	0.01	50	7.9	850
30	70	3	0	0	0.01	100	8.4	1240
76	30	16.1	0	+	0.01	25	8.1	730
Drinking water quality	Maximum 500	Maximum 3	0	0	Maximum 0.10	Maximum 45	6.5–8.5	Maximum 400*

MAB: mesophylls aerobic bacteria; TC: total coliforms; E. coli: Escherichia coli; PA: Pseudomonas aeruginosa; +, presence; 0, absence; pH: pH in water 1.2.5 and EC: electrical conductivity in water 1.2.5 by potentiometry [17].
References levels for drinking water quality according to CAA [41].
**of secondary importance according to CAA [41].*

Table 2.
Irrigation water characteristics.

productivity, lower product quality, higher disease incidence, higher energy use and lower efficiency in the use of water and fertilizers [42].

High pH levels in the irrigation water are due to the fact that it has sodium bicarbonate. When irrigating with this water, the sodium cation is concentrated in the soil, generating an alkaline reaction. Another problem, not less important, associated with the presence of sodium (Na) in the soil is the effect that this cation has on soil structure. Soils with high exchangeable sodium content have dispersed colloids and therefore an unstable structure. Moreover, soil crusts are formed that seal the surface, creating or magnifying the problems of fluid exchange and impedance for plants (seedling emergence) and biological activity. The accumulation of dispersing cations such as Na^+ causes expansion and/or dispersion of clays, which alters the geometry of the pores, in turn affecting the permeability and retention of water in the soil [38].

Studying soils from the southern green belt, a decrease in aggregate stability (AS) was found in cultivated soils, both under open-field and greenhouse cultivation systems, compared with uncultivated soils from house surroundings. Samples were analysed using the Le Bissonnais [16] technique that assesses structural stability by calculating the mean weight diameter (MWD) of soils aggregate afte r being subjected to three pre-treatment conditions to determine the dispersive effect of different processes. Thus, T1 aims to show how dry soil behaves in the face of heavy rain, T2 how wet soil behaves in the face of intense rain and T3 how dry soil behaves in the face of mild rain and therefore a slow humidification. Uncultivated soils presented higher aggregate stability after every pre-treatment, with the average index (AVE I) of aggregate stability of 2.47 for uncultivated soils and 1.1 for cultivated soils (**Figure 4**).

In addition, an increase in pH was observed in cultivated soils (**Figure 5**). **Table 3** shows the correlation found between aggregate stability and pH in soils from the southern green belt.

In the same study, when bulk density was analysed, no significant differences were found between cultivated and uncultivated soils (**Figure 6**). This is probably linked to the fact that the sampling was performed after soil tillage was carried

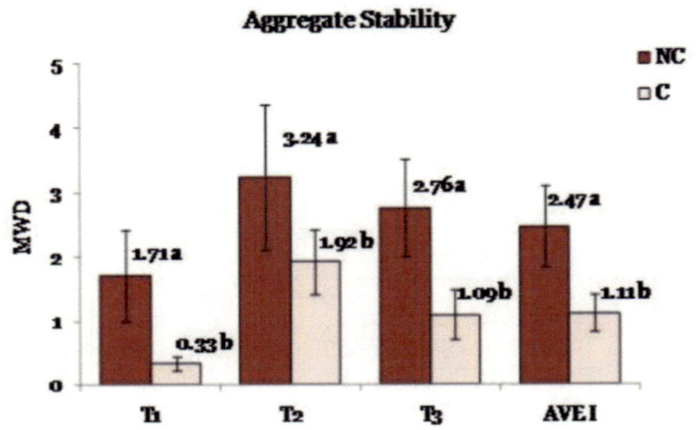

Figure 4.
Aggregate stability by Le Bissonnais method [16]. MWD: mean weight diameter; C: cultivation sites; NC: not cultivated; T1: dry soil behavior in the face of heavy rain; T2: wet soil behavior in the face of heavy rain; T3: dry soil behavior in the face of mild rain; AVE I: average index of the three pretreatments. Vertical bars indicate the standard deviation. Different letters show significant differences between treatments according to Tukey ($p < 0.05$) [33].

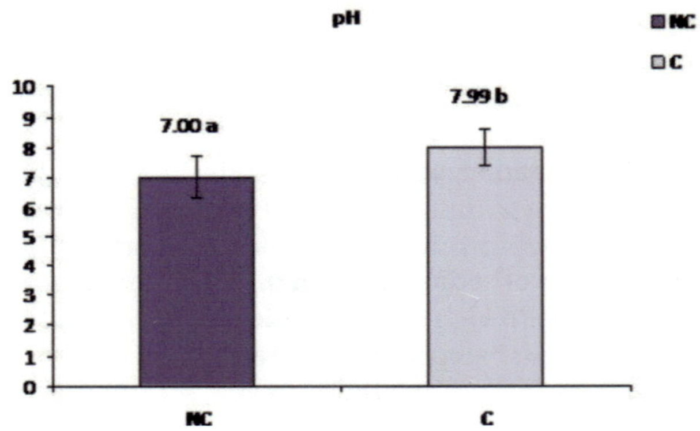

Figure 5.
Soils pH values. pH: pH in water 1: 2.5 by potentiometry [17]; C: cultivation; NC: not cultivated sites. Vertical bars indicate the standard deviation. Different letters show significant differences between treatments according to Tukey ($p < 0.05$) [33].

	T1	T2	T3	AVE I	pH
T1	1				
T2	0.66	1			
T3	0.91	0.73	1		
AVE I	0.95	0.79	0.97	1	
pH	−0.75	−0.44	−0.7	−0.7	1

T1: dry soil behavior in the face of heavy rain; T2: wet soil behavior in the face of heavy rain; T3: dry soil behavior in the face of mild rain; AVE I: average index of the three pre-treatments [16]; pH: pH in water 1.2.5 by potentiometry [16]. Significant correlations are presented with $p < 0.05$ [32].

Table 3.
Pearson correlation coefficients (r) between pH and aggregate stability treatments.

out in cultivated soils. Soil crusting and compaction due to structural instability leave producers with no options but to increase pre-sowing tillage to ensure crops emergence and growth. This causes greater aggregates destruction and formation of a plow floor that require re-tillage the soil generating a vicious circle that also increases production costs due to the high fuel use required for tillage tasks.

This instability of the production system causes, a few years after the start of production, the decrease in crop yields due to land degradation linked to the interconnection of the following problems: salinization, alkalization, decreased permeability, flooding, nutritional imbalances and development of diseases. To counteract the negative effects crop yields and without performing the appropriate analysis, fertilization doses are increased, adding soil hyperfertilization to the aforementioned problems [26] and increasing costs of production and environmental damage [37, 43].

It is common to try to reverse the decrease in yield by increasing the addition of fertilizers and phosphoric acid. Occasionally, there is a favorable initial response of the crop to greater fertilization, generating the wrong idea that it effectively contributes to improving the productive environment. However, eventually over fertilization occurs, and the problems associated with land degradation are exacerbated [21, 43] since these practices do not alleviate pH imbalances in a sustained manner and, on the contrary, they generate a negative effect due to the excessive accumulation of P (greater than 300 mg kg^{-1}) leading to nutritional imbalances and contamination [14, 43, 44]. With such high concentrations of P in the soil there is no response to fertilization [45] in addition, the availability of P not only depends on chemical fertilization, but also on the response or interaction with Ca^{+2}, Mg^{+2}, Al^{+3}, Fe^{+3} and Mn^{+2} [46]. Thus, in high pH situations, P forms insoluble compounds with Ca^{+2} and Mg^{+2} [45], leading to induced deficiencies. This has also been reported in other parts of the world with negative consequences on the environment, production and economics [47].

Figure 7 shows extractable P levels by Bray and Kurtz 1 found horticultural farms from the southern peri-urban. Sampling sites included soils under open-field and greenhouse cultivation systems and uncultivated soils adjacent to the houses. The range of extractable P concentrations varied from 50.7 to extremely high levels as 538.7 mg kg^{-1}, with an average of 255.0 mg kg^{-1}. It is also observed that the minimum levels obtained from cultivated sites double the maximum level from uncultivated sites. It is also interesting to note that the levels found in uncultivated sites are well above what soils should have in their natural state [9, 12, 34], so it could be hypothesized that such high P contents, would exceed soils retention capacity, therefore remaining in solution and being able to reach uncultivated sites from adjacent cultivated areas. In this sense, [48] reported contamination of a river basin in the southern peri-urban area with P derived from the horticultural activity.

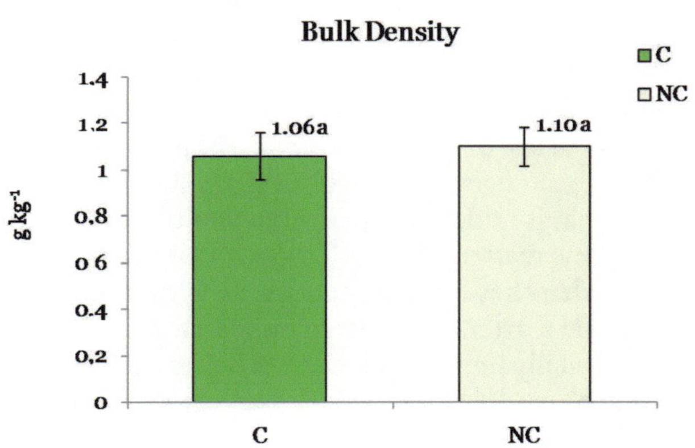

Figure 6.
Bulk density values. BD: bulk density by cylinder method [15]; C: cultivation sites; NC: not cultivated. Vertical bars indicate the standard deviation. Different letters show significant differences between treatments according to Tukey (p < 0.05) [33].

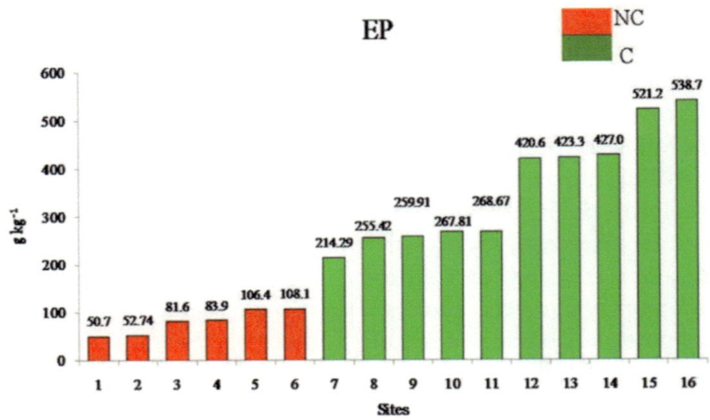

Figure 7.
Extractable phosphorus values in C: cultivated and NC: not cultivated sites. EP: extractable phosphorus by Bray and Kurtz 1 [20].

At present, there are few works that analyse the changes in the P content of the soil due to horticulture or that explain the causes of such accumulations of P in uncultivated sites in the area.

Another horticultural practice proposed as a possible solution to degradation soil but ended up generating major problems, is the use of poultry litter as organic fertilizer. This organic amendment is widely used in the area to improve the organic matter content of the soil and thus improve its structure; however, its excessive application is a potential source of soil and water contamination. In addition to providing nutrients, these fertilizers introduce salts and pollutants into the soil and, in general, these poultry litters are characterized by having high pH, high electrical conductivity and high P content, which contribute to worsen the aforementioned problems. These materials are very variable in their composition and therefore should be analysed prior to soil application. As an example, when analysing a poultry litter sample from a supplier from the area under study, pH levels found were between 7.6 and 8.6; electrical conductivity values between 3 and 7.8 dS m^{-1}, TOC between 34 and 45%, total N content between 2.08 and 2.42% and the C/N ratio between 16.3 and 18.6%. The extractable P content found ranged between 1600 and 3900 mg kg^{-1} and total P between 6500 and 12,000 mg kg^{-1}. Other authors have reported average total P content of 8700 mg kg^{-1} [25], although concentrations higher than 10,000 mg kg^{-1} have been detected in poultry litter and poultry manure [49, 50]. These characteristics depend on the composting time and the climatic conditions during composting. A poultry litter supplier usually sells to several horticultural farms of the area.

However, beyond all the negative characteristics of the poultry litter, some authors have proposed the use of these organic amendments as a promising practice to improve or mitigate the impact of horticultural production on the soil [51, 52]. An important process that is undergone in the horticultural soils under study is the progressive loss of organic matter [53]. [54] argue about the advantages of poultry residues and underline that they not only contribute with significant amounts of nutrients but also organic matter, which has a favorable effect on soils structure and permeability which is usually limited in the green belt area. However, other authors claim that the addition of poultry litter does not lead to significant increases in total oxidative carbon (TOC), since organic humus-forming materials are those of plant origin [26, 55]. The labile fractions of organic matter, such as those represented by the particulate oxidative carbon (POC), are much more sensitive to changes in management practices. Therefore, when a change in practices is made, the dynamics of these fractions are altered, affecting the physical, chemical and biochemical

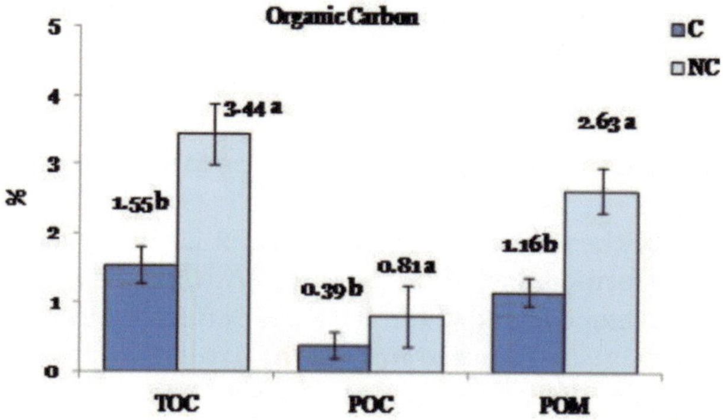

Figure 8.
Organic carbon fractions. TOC: total oxidative carbon by Walkley and Black [18]; MOC: oxidative carbon associated with the mineral fraction and POC: particulate oxidative carbon [19]. Vertical bars indicate the standard deviation. Different letters show significant differences between treatments according to Tukey ($p < 0.05$) [33].

environment of the soil [56]. Oxidative carbon associated with the mineral fraction (MOC) corresponds to a more complex and stable humus, which requires more time for its formation, making it difficult to increase its content through agricultural practices in the short term.

Figure 8 shows the results obtained from soils sampled in horticultural farms from the districts of La Plata and Florencio Varela where poultry litter is applied regularly every year, in some cases on more than one occasion. In the present study, the contents of TOC and MOC were lower in cultivated sites than in uncultivated sites. In addition, it should be noted that the POC was also lower in cultivated sites, indicating that the addition of poultry litter is not enough to counteract the loss of organic carbon associated with horticulture, not even those fractions of rapid cycling. Similar results were found by [8, 57].

4. Horticulture in the city

In the last years and in different parts of the world, the way the city is inhabited has been transformed; the daily relationship between human beings and their natural environment within the city has changed [58]. During the twentieth century, urban agriculture reached a great development due to increasing urbanization, deterioration of life conditions in poor neighborhoods, wars, natural disasters, environmental degradation and lack of resources, which caused food shortage. Urban agriculture means food production within the cities, in most cases it is a small-scale activity scattered throughout the city [59]. The large number of community horticultural gardens located in charitable dining rooms and in vacant spaces (for example, under high-voltage lines or along roads and waterways), or in institutional spaces such as hospitals and businesses, family gardens in backyards and roofs and school gardens are just a few examples that show the growing presence of agriculture in the cities [60].

Food production within the city is mainly used for self-consumption, to improve the amount of available food, for its freshness, variety and nutritional value [61], for environmental education and the exchange of experiences, among other factors, as [62] points out. It is also associated with jobs generation and income for groups of individuals, and it promotes environmental sanitation through the recycling of organic waste [63]. As it is a multifunctional and multicomponent activity, urban agriculture can respond to a great diversity of urban issues that include from the fight against poverty and the strengthening of self-esteem, to the improvement of the urban environment, participatory governance, management of the territory and food and nutritional security [60].

5. Front of organizations in fight (FOL) cooperatives: a case study

In Argentina, after the political-economic crisis of 2001, movements of unemployed workers were formed, including the FOL. Within these movements labor cooperatives were developed. One of the activities performed in a cooperative way is horticulture, which is carried out in community gardens located within popular neighborhoods, many of them developed in poor settlements. The majority of the production of these gardens is destined to the neighborhood dining rooms; therefore they are an important component improving the diet of hundreds of families, who otherwise only ingest carbohydrates and canned food. In turn, these spaces serve as a reference for the entire neighborhood and encourage exchange and knowledge between neighbors. Therefore, self-sustaining neighborhoods are generated with less garbage, different exchange relationships, and healthier food. However, horticultural gardens located in these popular neighborhoods often have one main problem; they were developed on soils highly modified. Urban soils or "anthropic soils" have been disturbed profoundly by human activity through the mixing, importing and exporting of materials [64, 65], and they are often characterized by contamination, compaction and soil sealing, as well as deposition, and removal or mixing of natural substrates. Soils in the urban environment tend to be very disturbed because of surrounding human activities and might even be exogenous (i.e., transported from elsewhere) [66, 67]. Therefore, the properties of urban soils are normally not favorable for plant growth and their role in food production is compromised [68].

Soils from horticultural gardens developed in poor settlements from La Plata city, were sampled during 2018 and 2019. Site 1 corresponds to "Barrio Altos de San Lorenzo," which is located on an old landfill pit that has received the contribution of the surrounding stream, Site 2 and 3 are located in "Barrio el Carmen," built on top of material dragged from the bottom of the Maldonado stream (**Figure 9**).

The most outstanding characteristics of these soils were: moderate bulk density, high pH and low TOC content (**Table 4**).

These conditions necessarily involve the application of soil recovery management practices in the horticultural gardens. In this sense, agroecology proposes soil management strategies that can contribute to improve their productive properties. To guarantee the success of the popular gardens within the same organizations, workshops and training for agroecological production are given. The gardens are fertilized with compost that is obtained from compostable household waste. Moreover, crop rotation and crops associations are considered to improve nutrient balances and aromatic species that serve as insect repellents are sown. Agroecology takes advantage of the natural processes of interactions that occur in an horticultural field in order to reduce external inputs (many of them potential contaminants and toxic compounds) and improve biological efficiency of cropping systems [69].

The horticultural gardens managed by FOL cooperatives in the neighborhoods of "Altos de San Lorenzo" and "El Carmen," unlike what happens in peri-urban productions, are irrigated with water from the water supply network which is of better quality than the water pumped from wells (lower electrical conductivity and lower pH). However, given the precarious nature of these settlements, many of them are illegally occupying fiscal lands, do not have legal access to the water supply network and make unsafe connections that can become contaminated with faecal matter (**Table 5**).

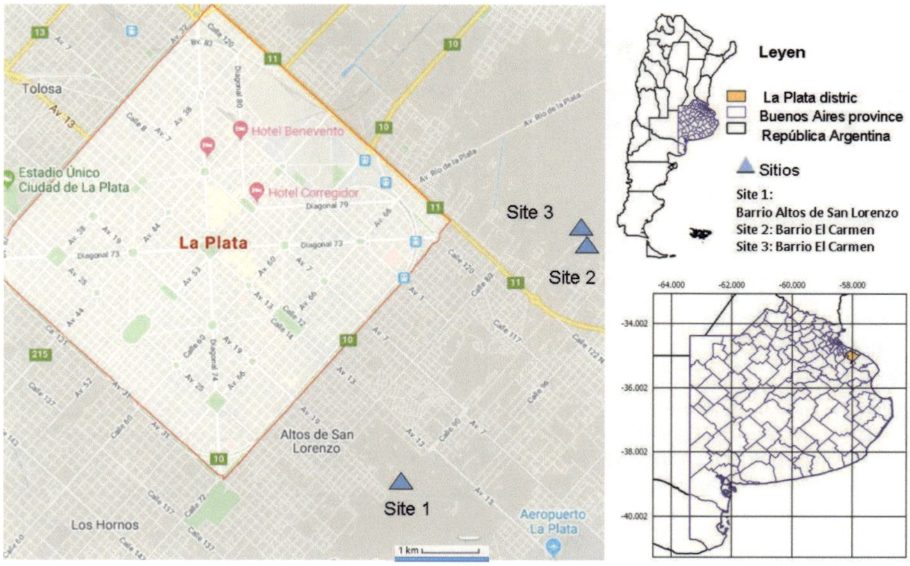

Figure 9.
Location of sampling sites.

Sites	BD g cm^{-3}	pH	TOC %	Texture			
				Sand %	Silt %	Clay %	Soil texture
Altos de San Lorenzo	1.17	7.91	0.74	27	34	39	Clay loam
El Carmen	1.28	8.70	1.14	30	28	52	Clay
Ideal scenario	Lower than 1.1	5.5–7.0	Greater than 3.5				

BD: bulk density by cylinder method [15]; Soil texture: clay, sand and silt percentage (USDA triangle); pH: pH in water 1.2.5 by potentiometry [17]; TOC: total oxidative carbon by Walkley and Black [18]. Ideal scenario: reference values for adequate crop production [62].

Table 4.
Urban soils characteristic in not cultivated sites: Altos de San Lorenzo y El Carmen.

After 3 years of agroecological production in the neighborhoods under study, soil sampling was performed and it was observed that some characteristics improved (**Figure 10**).

Bulk density diminished and an increase in TOC was observed (**Figure 10c** and **f**), these improvements are linked to the use of compost as an organic amendment in the agroecological gardens [70, 71] which not only acts as fertilizer but also improves soil physical properties [69].

Although urban soils from the popular neighborhoods studied had electric conductivity levels below the risk thresholds for any crop, this property decreased in the horticultural gardens which could be related to salts washing as a result of irrigation with non-saline water (**Figure 10b**). No improvements were found regarding total N content (**Figure 10d**), this could be explained due to the use of vegetable compost as the only fertilizer. Composting from household waste usually has low N content and high C/N ratio [72], therefore a source of nitrogen should be considered to improve soil quality, for example through the incorporation of green manure with legumes which are able to fix atmospheric nitrogen or including more

Site	Count		Investigation		Nitrites	Nitrates	pH	EC
	MAB u.f.c. ml^{-1}	TC NMP 100 ml^{-1}	*E. coli* in 100 ml	PA in 100 ml	ppm	ppm		µs cm^{-1}
Altos de San Lorenzo	100	9.2	+	+	0.01	30	6.90	973
El Carmen	Lower than 10	Lower than 3	0	0	0.01	10	6.95	1016
Drinking water quality	Maximum 500	Maximum 3	0	0	Maximum 0.10	Maximum 45	6.5–8.5	Maximum 400*

MAB: mesophylls aerobic bacteria; TC: total coliforms; E. coli: Escherichia coli; PA: Pseudomonas aeruginosa; +, presence; 0, absence; pH: pH in water 1.2.5 and EC: electrical conductivity in water 1.2.5 by potentiometry [17]. References levels for drinking water quality according to CAA [41].
**of secondary importance according to CAA [41].*

Table 5.
Irrigation water characteristics of urban sites.

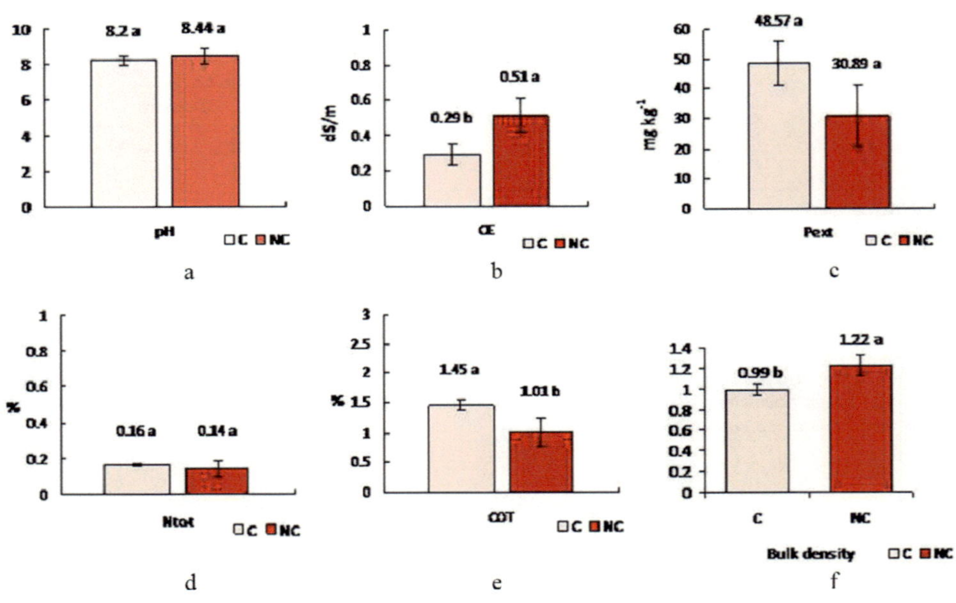

Figure 10.
(a) pH: pH in water 1; 2.5 (potentiometry) [17]; (b) EC: electrical conductivity in water 1; 2.5 (potentiometry) [17]; (c) TOC: total organic carbon (Walkley and Black) [18]; (d) TN: total nitrogen (Kjeldahl); (e) EP: extractable phosphorus (Bray and Kurtz 1) [20] and (f) BD: bulk density [15] for NC: not cultivated sites and cultivated sites (C). Vertical bars indicating the standard deviation. Different letters show significant differences between treatments according to Tukey ($p < 0.05$) [32].

legumes in crop rotation. Another property that could not be improved was pH (**Figure 10a**), which was expected, since pH is one of the chemical properties of the soil that varies the least, because it is an intrinsic characteristic of the soil genesis [73, 74]. To generate a significant change in pH, some specific corrective amendment should be applied, for example calcium sulfate.

With respect to extractable P [12, 34], reported that soils from this area, in their natural condition, are characterized by having low levels of extractable P (less than 10 mg kg^{-1}). Other studies in the southern peri-urban area reported higher values, close to 20 mg kg^{-1} in uncultivated soils [13, 14]. In this study, we worked with urban soils composed of material removed from the stream that receives runoff water from the entire basin. Phosphorus from many human activities, mainly the use of fertilizers and amendments in horticultural productions of the peri-urban area [8, 14, 26], could have been transported dissolved along with the runoff water and be retained in the clays [44] of the stream bed material that was subsequently used to build the urban soils of some settlements. This could be the reason why high levels of extractable P were found even in uncultivated soils from popular neighborhoods (**Figure 10e**).

Although this section aims to show how urban land can be improved for food production, it is important to take into account some sanitary characteristics of the material on which food is grown, such as knowing the content of heavy metals which are commonly raised in urban soils.

6. Conclusions

Horticultural soils from the green belt of Buenos Aires are showing alarming signs of physical, chemical and biological degradation as a consequence of inappropriate management practices applied since many decades ago. The most important processes associated with soil degradation in this area were salinization and alkalization principally as a consequence of irrigation with water with high levels of sodium bicarbonate and excessive application of organic amendments. The mentioned processes are also associated with nutritional imbalances and the loss of soil structure. Soil structure is also negatively affected by the loss of soil organic matter, usually observed in intensive agricultural systems as horticulture, which is not being compensated by organic amendments. On the contrary, the use of organic amendments and inorganic fertilizers indiscriminately, without performing the appropriate previous soil analyzes, or considering the needs of the crops, generates an over fertilization that increases the risks of nutritional deficiencies and could cause environmental damage due to nutrients leaching to underground water and superficial water courses. Particularly, there is a need for research on the dynamics of phosphorus since, although this element is considered immobile, the concentrations found in the soils of the area are so high that they could exceed the retention capacity of the soils and generate important environmental impacts at the basin level. Therefore, given the problems described in this chapter, it is necessary and urgent to change the productive paradigms if the intention is to ensure food production in the most important horticultural sector of the Argentine Republic. Soil is a non-renewable natural resource, its use and management must be integrated in a long-term perspective within a sustainable development approach; within a sustainable agriculture. Agroecology gives a new approach to the agricultural system, trying to provide solutions based on the interactions of physical, biological and socioeconomic components of the systems, integrating knowledge in the local and regional level to ensure sustainable production. Urban horticulture in Argentina is being developed mainly under an agro-ecological perspective. Although production within cities covers a much smaller area and of less economic importance than peri-urban horticulture, it is a role model, generating information from multiple local experiences that can serve as a basis to change large-scale horticultural production systems. The challenges that appear in urban production systems with an agro-ecological perspective are related, among other things, to the difficulties of producing in unnatural soils. In these soils, it is compulsory to perform quality analysis, treatment and transformation to ensure a healthy and sustainable production.

The search for agro-socioeconomical sustainability and new production system paradigms are the greatest challenges of modern agriculture, which involves among other technological practices, and adequate soil management.

Author details

Paladino Ileana, Sokolowski Ana Clara*, Prack Mc Cormick Barbara,
José Enrique Wolski and Rodríguez Hernán y Mauro Navas
Facultad de Ciencias Agrarias, Universidad Nacional de Lomas de Zamora,
Argentina

*Address all correspondence to: soko576@hotmail.com

References

[1] Fernández Lozano J. La Producción de Hortalizas en Argentina. Secretaría de Comercio Interior. Corporación del Mercado Central de Buenos Aires. Gerencia de Calidad y Tecnología; 2012. 29 p

[2] Leguizamón ES. Historia de la Horticultura. Ediciones INTA, Buenos Aires. 1ra. Edición; 2018. 49 p. ISBN: 978-987-521-927-4

[3] Di Pace M. Ecología de la ciudad. Colección Coedición 23. Editora Prometeo Libros, Buenos Aires: Prometeo; 2004

[4] Ministerio de Asuntos Agrarios. Censo Hortiflorícola de la Provincia de Buenos Aires. 2005. Dirección Provincial de Estadística (DPE) y Dirección de Economía Rural (DER); 2006

[5] Viteri ML, Ghezán G, Iglesias D. Tomate y lechuga: producción, comercialización y consumo. Proy Esp: Economía de las Cadenas Agroalimentarias y Agroindustriales Área Estratégica de Economía y Sociología Proyecto Propio de la Red Competitividad de las Cadenas Agroalimentarias y Agroindustriales. Estudio socioeconómico de los sistemas agroalimentarios y agroindustriales; 2013. ISSN: 1852-4605. N°14

[6] Gómez C, Mediavilla MC, Pineda C. Organizaciones de productores Bolivianos en el Cinturón Hortícola Platense y el desafío de alcanzar un modelo de producción cada vez más inclusivo y sustentable. In: Capítulo 11 en Migrantes bolivianos en el periurbano bonaerense. Memorias, producciones, políticas, trabajo, tecnología, organizaciones. Ediciones INTA, M.C. Feito Coordinadora; 2013

[7] Rivas I. Gestión ambiental para el ordenamiento territorial del Partido de Florencio Varela, Área Metropolitana de Buenos Aires. Revista de la Asociación Geológica Argentina. 2010;**66**(4):535-543

[8] Cuellas MV. Horticultura Periurbana: Análisis de la fertilidad de los suelos en invernaderos. Chilean Journal of Agricultural & Animal Science. 2017;**33**(2):163-173

[9] Balcaza LF. Deterioro de los suelos bajo invernáculo. Revista IDIA XXI Sección Floricultura; 2003. pp. 196-200

[10] Soil Survey Staff. Key to Soil Taxonomy. Twelfth ed. United States Department of Agriculture, National Resources Conservation Service; 2014

[11] World reference base for soil resources, 2015. In: World Soils Resources Reports, N° 106. 191 pp. Roma: FAO; 2015

[12] Hurtado M, Giménez J, Cabral M. Análisis ambiental del partido de La Plata: Aportes al ordenamiento territorial. La Plata, Buenos Aires, Argentina: Consejo Federal de Inversiones; 2006. 134 pp

[13] Andreau R, Gelati P, Provaza M, Bennardi D, Fernández D, Vázquez M. Degradación física y química de dos suelos del cordón hortícola platense. Alternativas de tratamiento. Cs Suelo. 2012;**30**(2):107-117

[14] Paladino IR, Sokolowski AC, Irigoin J, Rodriguez H, Gagey MC, Barrios MB, et al. Soil properties evaluation in horticultural farms of Florencio Varela, Buenos Aires, Argentina. Environment and Earth Science. 2018;**77**:411

[15] Blake GR, Hartge KH. Bulk density. In: Klute: Methods of Soil Analysis. 2° ed. Vol. 9. Madison: American Society of Agronomy (ASA); 1986. pp. 363-375

[16] Le Bissonnais Y. Aggregate stability and assessment of soil crustability and erodability. I. Theory and metodology. European Journal of Soil Science. 1996;**47**:425-437

[17] SAMLA. Secretaría de Agricultura, Ganadería, Pesca y Alimentación de la Nación Argentina. Dirección de Producción Agrícola. Sistema de Apoyo Metodológico a los Laboratorios de Análisis de Suelos Agua, Vegetales y Enmiendas Orgánicas. Buenos Aires, Argentina; 2004. 120 pp

[18] Jackson ML. Análisis químicos de suelo. Ediciones OMEGA, S.A. Casanova 220 Barcelona Tercera Edición; 1976

[19] Cambardella MR, Elliot ET. Particulate organic matter changes across a grassland cultivation sequence. Soil Science Society of America Journal. 1992;**56**:777-783

[20] NORMA IRAM-SAGyP 29570-1. Norma Argentina. Calidad ambiental y calidad de suelo. Determinaciones de P extraíble de los suelos. 1era edición; 2010. 20 pp

[21] Alconada M, Zembo J, Mortola N. Influencia cualitativa del riego con aguas subterráneas en suelos con producciones intensivas a campo y en invernáculo. 1er Joint World Congress on Groundwater, Brasil; 2000

[22] Abril F, De Paz JM, Ramos C. El riesgo de lixiviación de nitrato en las principales zonas hortícolas de la comunidad valenciana. Jornadas sobre la contaminación de las aguas subterráneas: un problema pendiente. Valencia; 1998. pp. 65-71

[23] Polack LA. Tecnología apropiada para la sustentabilidad de sistemas hortíflorícolas con énfasis en cultivos protegidos. In: PTR 2013-2019. Instituto Nacional de Tecnología Agropecuaria (INTA): Buenos Aires; 2013

[24] INTAGRI. Propiedades Físicas del Suelo y el Crecimiento de las Plantas. Serie Suelos. Núm. 29. Artículos Técnicos de INTAGRI. México; 2017. 5 p

[25] Alconada M. Desinfección del suelo con vapor. Efectos sobre la nutrición de los cultivos. Argentina: Ediciones, INTA; 2004. 123 p

[26] Alconada M, Cuellas M, Poncetta P, Barragán S, Inda E, Mitidieri A. Fertirrigación en un cultivo de tomate protegido: I-Nutrición nitrogenada. Efectos en el suelo y en la producción. Revista Horticultura Argentina. 2011;**30**(72):5-13

[27] Poncetta P, Alconada M, Lavado R. Producción de tomate protegido en suelos decapitados con diferentes planes de fertirrigación. XX Congreso Argentino de la 681 Ciencia del Suelo, Salta. Nro. 481; 2006

[28] Ostarrieth M. Gira de campo. XVII Congreso Argentino de la Ciencia del Suelo. Mar del Plata; 2000

[29] Magallón J, Tremont O, Rodríguez N. Efectos del uso de un vermicompost sobre las propiedades biológicas y químicas de los suelos degradados por sales. Venesuelos. 2001;**9**(1-2):48-56

[30] Cabrera A, Arzuaga J and Mojena M. Desbalance nutrimental del suelo y efecto 601 sobre el rendimiento de tomate (*Lycopersicum solanum* L.) y pepino (*Cucumis sativus* 602 L.) en condiciones de cultivo protegido. Cultivos Tropicales. 2007;**28**(3):91-97

[31] Pérez Vázquez A, Landeros SC. Agricultura y deterioro ambiental. Elementos. 2009;**73**(16):19

[32] Szczesny A, Adlercreutz E, Huarte D, López Camelo A, Manzo E, Viglianchino L. Producción hortícola bajo cubierta. 150 p. Argentina: Instituto Nacional de Tecnología Agropecuaria

(INTA), Ciudad Autónoma de Buenos Aires; 2014. p. 150

[33] De Rienzo A, Casanoves F, Balzarini M, González L, Tablada M, Robledo W. Infostat software estadístico. Argentina: Grupo Infostat, FCA, Universidad Nacional de Córdoba; 2008

[34] Balcaza LF. Fertilización de pimiento. IDIA XXI. 2003a;(4):114-120

[35] Nuez F, Ortega G, Costa J. El cultivo de pimientos, chiles y ajíes. Madrid, España: Mundi Prensa; 2003. p. 607

[36] Cadahia LC. Fertirrigación: Cultivos hortícolas y ornamentales. 3ra ed. Madrid, España: Mundi Prensa; 2005. p. 681

[37] Alconada M and Huergo L. 1998. Degradación de suelos con cultivos protegidos: Tomate. Influencia de la calidad de agua de riego. I Reunión de Producción Vegetal, N.O.A. Universidad Nacional de Tucumán, Versión electrónica, Argentina

[38] Porta J, López Acevedo M, Roquero C. Edafología para la agricultura y el medio ambiente. 3ra ed. Madrid, España: Mundi Prensa; 2003. p. 930

[39] Maroto JV. Elementos de horticultura general. Madrid, España: Ed. Mundi Prensa; 1990. ISBN: 84-7114-297-X

[40] Kang Y, Park J, Kim S, Kang N, Park K, Lee S. Effects of root zone pH and nutrient concentration on the growth and nutrient uptake of tomato seedlings. Journal of Plant Nutrition. 2011;**34**:640-652

[41] Código Alimentario Argentino. Capítulo XII Bebidas Alcohólicas. Bebidas hídricas, agua y agua gasificada. Agua potable. Art. 932; 2008

[42] Lang M, Lipinski V, Balcaza LF. Efecto de la aplicación de enmiendas y manejo del riego sobre el rendimiento y la calidad de lechuga bajo invernadero. Horticul Argent. 2013;**32**(79):32-43

[43] Giuffré L, Alconada M, Pascale C, Ratto S. Environmental impact of phosphorus overfertilization in tomato greenhouse production. Journal of Applied Horticultre. 2004;**6**(1):58-61

[44] Vázquez M, Terminiello A. Recuperación de suelos degradados de pequeños productores del cinturón hortícola del Gran La Plata. Valoración del problema y estrategias correctivas. Argentina: FCAyF, UNLP; 2008

[45] Balemi T, Negisho L. Management of soil phosphorus and plant adaptation mechanisms to phosphorus stress for sustainable crop production: A review. Journal of Soil Science and Plant Nutrition. 2012;**12**:547-562

[46] Carrijo O, Hochmuth G. Tomato responses to preplant-incorporated or fertigated phosphorus on soils varying in Mehlich-1 extractable phosphorus. HortScience. 2000;**35**(10):21273

[47] Karami E, Ebrahimi H. Overfertilization with phosphorus in Iran: A sustainability problem. Journal of Extension Systems. 2000;**16**:100-120

[48] Delgado MI, Carol E and Tanjal C. 2018. Flujo de nutrientes asociado a la relación agua superficial-subterránea en una cuenca con creciente desarrollo de Agricultura intensiva. In the Book El Agua Subterránea: Recurso sin Fronteras: Interacción Agua Superficiales y Agua Subterráneas. Hidrogeología de Salares. Primera Edición. Editorial de la Universidad Nacional de Salta, Salta

[49] Leconte MC. Efecto del agregado de estiércol de gallina en el compostaje de aserrín y cascarilla de arroz. Doctorado de la Universidad Nacional del Nordeste

Area Recursos Naturales Facultad de Ciencias Agrarias. Argentina: Universidad Nacional del Comahue; 2010. 199 pp

[50] Barbazán M, del Pino A, Moltini C, Hernández J, Rodríguez J. Caracterización de materiales orgánicos aplicados en sistemas agrícolas intensivos de Uruguay. Agrociencia Uruguay. 2011;**15**(1):82-92

[51] Lal R. Carbon sequestration in soil. CAB reviews. Perspectives in Agriculture, Veterinary Science, Nutrition and Natural Resources. 2008;**3**:10.1079

[52] Coscia P, del Pino A, Barros C and Moltini C. Cambios en la materia de suelos hortícolas de Uruguay con agregado de enmiendas orgánicas (CD ROM) En: Reunión Técnica: Dinámica de las propiedades del suelo bajo diferentes usos y manejos. Colonia; 2010

[53] Comese RV, González MG, Conti ME. Cambios en las propiedades de suelo de huerta y rendimiento de Beta vulgaris var. cicla por el uso de enmiendas orgánicas. Ciencia del Suelo. 2009;**27**(2):271-275

[54] Rostagno HS, Dionizio M, Paez LE, Butery CB, Albino LF. Impacto de la nutrición de pollos de engorde sobre el medio ambiente. Bolivia: Memorias XVIII Congreso Latinoamericano de la Avicultura; 2003. p. 431

[55] Labrador Moreno J. La materia orgánica en los agrosistemas. Edi Mapamundi-prensa Madrid; 1996

[56] Six J, Elliot ET, Paustian K. Aggregate and soil organic matter dynamics under conventional and no-tillage systems. Soil Science Society of America Journal. 1999;**63**:1350-1358

[57] Calandrelli L, Falcón ML. Efecto de enmiendas orgánicas sobre el suelo y cultivo de lechuga protegido, Trabajo final de carrera. Argentina: UNLP; 2018

[58] Millan Ortiz CP. Historia de la agricultura urbana y periurbana en la localidad Santa Fe (2004-2016). Trabajo de Grado para Optar al Título de: Licenciada en Ciencias Sociales Universidad Pedagógica Nacional Facultad de Humanidades Departamento de Ciencias Sociales Bogotá DC., Colombia; 2017

[59] Hernández L. La agricultura urbana y caracterización de sus sistemas productivos y sociales, como vía para la seguridad alimentaria en nuestras ciudades. Cultivos Tropicales. 2006;**27**(2):13-25. Instituto Nacional de Ciencias Agrícolas La Habana, Cuba

[60] FAO. Ciudades más verdes en America Latina y el Caribe. Informe sobre agricultura urbana y periurbana en la región. E-ISBN: 978-925-308251-3 PDF; 2014 (Available from: www.fao.org/ publications)

[61] Chavarrías M. Agricultura Urbana y Seguridad Alimentaría. Diario de la Seguridad Alimentaria; 2005

[62] Zaar MH. Agricultura Urbana: Algunas Reflexiones sobre su Origen y Expansión. Biblio 3W. Revista Bibliográfica de Geografía y Ciencias Sociales, XVI. Universidad de Barcelona: Barcelona; 2011

[63] Figueroa J, Izquierdo J. Agricultura Urbana en la región Metropolitana de Santiago de Chile: Situación de las Empresas Familiares Hidropónicas-estudios de casos. Santiago Chile: Oficina Regional de la FAO para América Latina y el Caribe; 2002

[64] De Kimpe CR, Morel JL. Urban soil management: A growing concern. Soil Science. 2000;**165**:31-40

[65] Pereyra FX, Paladino IR, Villegas D, Patitucci M. Nueva Propuesta de Clasificación de Suelos urbanos en Zonas Antropizadas. Tucumán: Actas XX Congreso Geológico Argentino; 2017

[66] Craul PJ. Urban Soils: Applications and Practices. New York: John Wiley and Sons, Inc.; 1999

[67] Bullock P, Gregory PJ. Soils in the Urban Environment. Chichester: John Wiley and Sons; 2009

[68] Lal R, Stewart BA, editors. Urban Soils. Series: Advances in Soil Science. New York: CRC Press; 2017. pp. 406

[69] Sarandón SJ, Flores CC, editors. Agroecología: bases teóricas para el diseño y manejo de Agroecosistemas sustentables. Editorial de la UNLP; 2014

[70] Echiburú AA, Jorquera LI. Guía práctica campesina: Prácticas agroecológicas para mejorar la huerta familiar. Chile: Programa de las Naciones Unidas para el Desarrollo; 2016. ISBN: 978-956-7469-74-1

[71] Tannfeld JR. La huerta agroecológica de autosustento. INTA Agencia de Extensión Rural Sáenz Peña—Programa ProHuerta; 2011

[72] Polo MG, Kowaljow E, Castan E. Persistent effect of organic matter pulse on a sandy soil of semiarid Patagonia. Biology and Fertility of Soils. 2014;**51**(2):241-249

[73] Shi Z, Wang K, Bailey JS, Jordan C, Higgins AH. Temporal changes in the spatial distributions of some soil properties on a temperate grassland site. Soil Use and Management. 2002;**4**:353-362

[74] Cox MS, Gerard PD, Melinda AJ. Selected soil properties variability and their relationships with yield in three Mississippi fields. Soil Science. 2006;**171**:541-551

Chapter 9

Historical Gardens as an Inspiration for the Future of Urban Horticultural Gardens

Ines Babnik

Abstract

Throughout the history people incorporated designed gardens in their closest living environment. They shaped their environment in such a way as to make it more useful, pleasing, and nicer. The old ancient civilization already created gardens that amazed anyone visiting the city—a good example are the great cities of Mesopotamia with hanging gardens and city entrance gardens dedicated to flowers, shrubs, and trees, creating a feeling of being in paradise. Renaissance gardens brought a great diversity of new garden motifs and innovations, while Baroque gardens presented the whole city in themselves, creating green walls and green architecture. The nineteenth century with its industrial revolution offered new technologies, new ways of designing and adjusting the nature to man's need. The twentieth and twenty-first centuries brought to us various ways to include green elements ranging from small to large-scale in our living environment, (from greenhouses in the parks to green walls inside the buildings). Through different motifs of historical gardens, we can find possibilities for today's and future urban horticultural gardens.

Keywords: historical garden, urban horticulture, prospects, water motif, green wall, pot cultivation, art projects, Baroque garden, Renaissance garden

1. Introduction

We humans are very practical creatures. We modify our surroundings to suit our needs—thus, we have been reshaping nature so that it would serve us best in a utilitarian and/or aesthetic sense. Throughout the history, a variety of gardens has been created—historical gardens (preserved or merely written about)—that today can give us a good insight into how resourceful humans were in a particular period of history or even provide us with ideas for our own living environment. Today, the term urban horticulture has become impossible to overlook—as stated by the United Nations: "today, 55% of the world's population lives in urban areas, a proportion that is expected to increase to 68% by 2050" [1]; therefore, it is even more appropriate to look back—just to see the future more clearly.

The definition of horticulture emphasizes the scientific and artistic way of managing plants with the goal of obtaining food and different materials or providing comfort and decoration. We can trace the origin of horticulture back to ancient civilizations—the Persians were great experts in this field. As Relf [2] nicely pointed out when interpreting the definition of horticulture as a synthesis of plants and humans, horticulture "encompasses PLANTS, including the multitude of products (food, medicine, O_2) essential for human survival; and PEOPLE, whose active and passive involvement with 'the garden' brings about benefits to them as individuals and to the communities and cultures they comprise." Humans and plants are therefore an essential part of horticulture. It was man's desire to take a particular plant from its natural environment and integrate it into the environment close to his home, which led to the emergence of designed gardens. And the idea of a designed garden could only be born when the people's goal was no longer survival and when the individual had free time and energy to beautify his or her surroundings [3].

Gardens have grown over time, as human knowledge has grown (in the fields of horticulture, mechanics, construction, etc.), and today historic gardens are a wonderful treasure trove of examples and ideas of how humans once incorporated nature into their living environment and how they can do that today or in the future. In this chapter, we will look at some examples from the history of garden design, and through these we will try to present some possibilities for future urban horticultural gardens.

2. A variety of water features

Among most popular garden motifs are water motifs. While these require mostly engineering skills, knowledge of aquatic plants, including their specificities, and requirements is an important part when designing a water motif. The Renaissance brought a real wealth of water motifs which were further enhanced by the Baroque. Renaissance cascades,[1] for example, at Villa Lante in Italy, and water jets splashing out of sculptures or directly from the water surface, for example, at Villa d'Este in Italy (e.g., **Figure 1**), were common garden features which later, in the Baroque period, grew in magnitude, as evidenced by, for example, the cascades in the German Kassel (e.g., **Figure 2**) or the pompous fountains with ruler iconography in Versailles (e.g., **Figure 3**; [4, 5]). However, even if these motifs seem to be suitable only for aristocratic gardens and are a remnant of past ages, contemporary landscape architectural projects indicate the opposite. Namely, such historic examples

Figure 1.
The One Hundred Fountain (Le Centro Fontane) at the Villa d'Este, Tivoli (near Rome), Italy.

[1] Cascades are a water motif in a garden where water slowly flows (usually down a steep terrain) into lower basins. Such water "steps" can be natural or man-made. In gardens, these motifs were popular for their sound effect. They were common in sixteenth-century Italian gardens and from there spread into French gardens. There are beautiful examples in the gardens in and around Rome (the Villa of Lante, the Villa of Aldobrandini, the Villa d'Este, etc.), which are also rich in sculptural motifs [6, 7].

Figure 2.
Herkules with Oktogon and Großen Kaskaden, Kassel-Wilhelmshöhe, Germany.

Figure 3.
Bassin du char d'Apollon, fountain in the Parc de Versailles, France.

of water motifs were a useful source from which masters, such as the American landscape architect, designer, and teacher Lawrence Halprin (1916–2009), have drawn their ideas. Halprin created several recognizable water features with cascades in which the art of Renaissance and Baroque as well as the art of unspoiled of nature are combined (e.g., Franklin Delano Roosevelt Memorial in Washington DC from 1997). Today, water has also entered urban areas in such a way that it is no longer clearly separated from its surroundings—as was the case with water motifs in historical gardens. Today water is a part of the surface on which the user of the garden (or open public space) walks; it has crossed the borders and become a part of public surfaces. An example is the water motif above Ross's Landing Riverfront Park in Chattanooga, Tennessee, where water flows down multiple levels, connecting the city and the river. Similarly, high water jets, the most prominent element of Baroque fountains, are today merged with town squares and offer playgrounds to children and adults, allowing them at least to cool down on hot days (we could find examples all over the world—let us only mention Smale Riverfront Park, Cincinnati, whose planning stared in 1997 [8], and Viertel Zwei in Vienna whose construction started in 2007 and where the water jets are placed in a small square connecting newly built apartment buildings and service facilities (e.g., **Figure 4**). Among the variety of water motifs in today's cities, we can also find dry motifs that turn into water motifs only when water (mostly rain) is provided. An example is a canal on a narrow medieval street in Ljubljana, Slovenia, where small sculptures of a prominent Slovenian sculptor Jakob Brdar are placed in a canal and a vertical pedestal, also marked by

Figure 4.
Water jets in Viertel Zwei (after 2007), Vienna, Austria.

the sculptor's work, points to the change in horizontal structure of the street. When it starts raining, the canal is filled with water and the sculptures look like they are swimming in the canal (e.g., **Figure 5**).

The abovementioned water motives include waterfalls or spurts of water in fountains, but in historical gardens there where also calm surfaces of water reflecting the sky and sun and the objects near the water, usually emphasizing their meaning. Narrow or wide canals or smaller and larger pools were initially meant to provide water for the gardens. Thus, ancient civilization used them as part of their irrigation systems. However, in the New Ages, the canals and pools provided other uses closely tied with symbolic meaning. The best example is king's garden in Versailles, the gardens of monarch Louis XIV, who wanted to be presented as an absolute monarch, untouchable and distant, as the Sun King, and to demonstrate his absolute power, even over nature. There is an abundance of motives filled with symbolic meaning in the Versailles garden, but let us look at the water parterre composed of two large pools near the Versailles castle. Those pools had a very practical purpose, as well as a symbolical one. The pools reflect the sun's rays and light up the outside wall of the Hall of Mirrors, bringing the light

Figure 5.
Architectural biro Medprostor and sculptor Jakob Brdar (project realization in 2014), Ljubljana, Slovenia.

also inside, increasing the lightness of the Hall. One of the pools was decorated with sculptures representing male allegorical figures of four main rivers in France, emphasizing the greatness of the king's territory. The magnificence of the ruler was celebrated also in the grand canal which was about a mile long; it was used for naval demonstrations and had gondolas donated by the Republic of Venice, steered by gondoliers. Furthermore, large water surfaces that provided space for such demonstrations (water battles and rides with gondolas) were not so rare, - they could also be found in the Baroque king's gardens in Hanover and München (Germany; [4, 9–11]). The reflective quality of the still water that doubled the presence, beauty, or power of the surrounding objects and also gave an observer a second window to what he/she gazed upon (calling into question the limits of the present world) was popular in Baroque and Rococo gardens (e.g., **Figure 6**). In the late seventeenth and eighteenth centuries—with the new English landscape garden—calm water surfaces gain new role. They become a part of an idyllic pastoral landscape that the new garden style aimed to create. Lakes and ponds in the garden were walked around or crossed over. They had natural shapes and in their vicinity there was usually a pavilion or some other smaller architectural object. Their main aim was to create a romantic, even sentimental atmosphere, to bring tranquility to to the garden's users and to create a picturesque scenery for walkers to enjoy with each step they made. This role of the lake or pond that recreated a part of natural scenery in a human made garden was transferred into cities. It became a part of the human quest to bring nature into the city. A good example is Central Park in New York, where the landscape architect Frederick Law Olmsted and the designer Calvert Vaux created a city park with lakes in 1857 (completed in 1876). Today lakes and ponds are part of numerous city parks. When a new neighborhood that includes green designed spaces is planned within a city, such lakes and ponds are often part of the built area. A nice exampl is Viertel Zwei in Vienna, where a lake

Figure 6.
Water canal in Rococo garden and summer residence Sanssouci in Potsdam, Germany.

is the central point of the open space between the business buildings, providing a calming view through the window and a soothing atmosphere for lunch breaks (e.g., **Figure** 7; [12]).

Water is an important element in the human environment. It has always been attractive to people, not only because of the necessity of survival but also because of the cold, humidity, relaxation, and play that it offers during the hot months. Water is invigorating, not only visually and haptically but also in an auditory sense. The murmur of water inspired the old masters to seek ways to give the water even more voice. To this end, hydraulis, an organ-like machine, was created in antiquity. It was a manually operated machine. During the Renaissance, which is certainly considered to be the most innovative era in the history (especially the garden history) of the western world, the so-called hydropneumatic automatophone was created. The beginnings of this Renaissance invention date back to around 1550 at the aforementioned Villa d'Este, where Cardinal Ippolito II d'Este (1509–1572) created a magnificent Renaissance garden with numerous water features: 500 liters of water per second passed through 51 wells, 364 fountains, 220 pools, and other water motifs. Pirro Ligorio, who created the garden, gathered his knowledge by examining the nearby Hadrian's Villa. The creators and the garden owner himself also drew knowledge from older literature, e.g. Vitruvius' *De architectura libri*

Figure 7.
Lake in Viertel Zwei (after 2007), Vienna, Austria.

Figure 8.
Fontana dell'Organo (when no water is running, the organs are visible), Villa d'Este, Tivoli, Italy.

Figure 9.
Nikola Bašić, Sea organs, Zadar, Croatia.

decem, which was translated into Italian in 1567 and contains chapters on hydraulic installations. In the garden of the Villa d'Este stood the Fontana dell'Organo (organs fountain; e.g., **Figure 8**) and Fontana della Civetta (owl fountain), made by the French engineer Luc Leclerc and his nephew Claude Vernard. Both fountains created special sound effects for the visitors to enjoy, imitating the birds chirping (at the owl's fountain) or the sound of water organs (at the organs fountain), accompanied by richly decorated surroundings. The Fontana dell'Organo (consecrated in 1571) had a sophisticated mechanism hidden behind the fountain. Water was routed through pipes, canals, across a wheel, through smaller pools—sounds were produced by both water and displaced air. The visitors who came to see and listen to those fountains could witness a real concert [13]. These water instruments were popular in antiquity and afterwards in the early modern times, and they persist; even in modern times they have not been forgotten. In Zadar, Croatia, sea organs designed by architect Nikola Bašić were presented in 2005 as part of the Old Town Coast restaurant project (e.g., **Figure 9**). Sea organs are an architectural object and at the same time an experimental musical instrument. They are designed as a group of pipes which are placed under large marble steps leading to the sea and are hidden from the view of the observer (as was the case with aforementioned Renaissance organs). The waves hitting the pipes and filling them with water create air pressure that produces random but at the same time harmonious tones in the pipes.

This luxury of water motifs, which are already successfully integrated into urban tissues offers many opportunities for urban horticulture. Irrigation systems, or even more complex food production systems, such as those found in aquaponics, can include water cannons, cascades, walk-on water surfaces, and even water organs—hydropneumatic instruments.

3. Indoor

Closely related to water is another garden motif created in the Renaissance—the *grotta*.[2] In built cavities, garden visitors and users were able to escape from

[2] The grotta (It.) is an artificial cavity or a cave that mimics natural cavities. The grotta can be artistically sophisticated (such as nymphaeum), or it may have humorous motifs, or be focused on sound effects. It was already popular in the Renaissance and even more so in the time of Mannerism. It was first mentioned by Leon Battista Alberti. With its humid and cool climate, it is a pleasant place to stay in the summer. Grottas are usually decorated with shells, snails, pebbles, and minerals (oysters, pearls, brass, tuff, colored enamel, etc.; [6, 7]).

the heat or hustle and bustle of events in the garden, to cool themselves down, marvel at the sounds of water and wind, read stories embodied in sculptural decoration, and chase the glare of water on artificial cave walls, often claded with shells. For urban horticulture, these artificial caves could be quite interesting because it is not difficult to integrate them into the built structures of cities. They could find their place in a small square, on the roofs of apartment houses, or in steep slopes of riverbeds. A grotta has its own closed water system and can even allow plants to thrive in its wet and humid environment. In addition, in Baroque gardens the grotta was often a part of an architectural structure that provided space for an artificial cave in the lower part, while in the upper part one could find a pavilion, a festive hall, or a kind of viewpoint. Here, the visitors liked to linger, enjoy the view, and listen to the sounds of water coming from the grotta. Often, the sound of the water was also used as an accompaniment to small concerts on the upper floor. Such two-story garden facilities in historic gardens were also intended for other functions (the range of garden houses and architectural scenery in gardens expanded in the late eighteenth century with the English landscape style); from banquets, events, spectacles, games, music pavilions to guest rooms in the upper floor and to retreat rooms, artisan workshops, or cold stores in the lower part. The lower part of such buildings was often dug into the slopes, thus providing a constant temperature. Already in the Baroque period (or since the time of Catherine de'Medici in the sixteenth century when ice cream was invented), the nobility liked to place cold stores in the vicinity of their mansions. Ice was brought into these cold rooms in the winter, accumulated in large quantities in the middle of a usually centrally designed room, where it was preserved and helped refrigerate food for as long as 6 months [9, 14, 15]. Would such a cold store also benefit a modern city in the time of global warming (instead of electronic devices which, when cooling, also emit large amounts of heat)?

4. Green verticals

Among the more prominent and enthusiastically accepted projects of urban horticulture are certainly green walls, also called living walls or even vertical gardens. After the botanist Patrick Blanc created his first successful large indoor

Figure 10.
Patrick Blanc (with architect Jean Nouvel), an outdoor green wall, Musée du quai Branly, Paris.

green wall in 1986 (Cité des Sciences et de l'Industrie in Paris), these structures started springing up indoors and outdoors, in small or large scales, monocultural or mixed, creating patterns, images, or just a pleasant green "screen" of plants (e.g., **Figure 10**). Origins of green walls can be found already in the hanging gardens of Babylon, however, it seems more plausible that the idea and form of green walls stem from the green walls of Baroque gardens—the *bosquett*.[3] In Baroque, the typical garden design was based on regular, geometric patterns. The so-called formal garden design emphasized the rational supremacy of man over nature—this was reflected not only in the floor plan of the garden but also in the plants. These were sheared and shaped into certain straight lines, images of things, animals, or even human-like shapes. This art of shearing and growing plants was called *ars topiarium*. It was used by the old Greeks and Romans and revived in the Renaissance. Thus, in Baroque, garden designers continued and upgraded this art of plant designing by creating long green walls, corridors, or even streets with high and sheared trees and shrubs (e.g., **Figure 11**). Green borders led the view into infinity, giving a sense of the grandeur of nature controlled by man as well as offering a retreat into the green spaces behind the green walls. There were smaller spaces of a more intimate nature inside such bosquetts—such rooms contained benches, wells, or even additional green architectures (e.g., treillage).[4] Today's green walls are no different from those of Baroque gardens: when observing tall buildings clad in green walls, they appear as green corridors in Baroque gardens along which gentlemen walked after they left the parterre.

Figure 11.
Bosquett in Versailles, France.

[3] Bosquett (Fr.) is a wooded part of a designed garden or a carefully groomed woodland in garden designs. High sheared hedges create special spaces—green rooms, cabinets (Gartenraume, nem.; Salle de verdure, Fr.). These can have different functions and motifs (corridor, a ballroom, a pond, a lounge, an amphitheater, etc.). In Baroque gardens, bosquett was most often arranged behind the parterre. It was popular in the mid-seventeenth to mid-eighteenth centuries in major formal garden designs [6, 7].

[4] Treillage (Fr.) is a green corridor consisting of a wooden or steel frame on which the plants climb—it has an architectural character. Most often the frame is made of green- or white-painted wood. These green hallways have pavilions at their intersections or corners. They offer walks and contemplation in the shade. Trellis (Fr.) is merely a framework for climbing plants; it is simpler and smaller in size than treillage [6, 7].

5. Ornamental parterre

The parterre[5] is another interesting element of historic gardens. In Baroque gardens, parterres were usually arranged next to the mansion. Thus, the first (or second) floor of the mansion offered the most beautiful view of the parterre. Parterres featured different colors, materials, and patterns. Their appearance varied throughout the year—to keep up with vegetative seasons, gardeners needed to quickly change the plants. Of course, there were also parterres (mainly Renaissance ones) composed of only box trees, sand (of different colors), and grass. However, in the seventeenth century, parterres that included diverse selection of flowers became more numerous. To allow the flowering pattern to be changed quickly and efficiently (to replace color, height, texture of the plant, etc.), the plants were often planted in pots (pot gardening). Thus, the plants no longer in bloom were easily replaced with the then flowering plants. This kind of gardening practice is still used today except that we do not put the containers into holes in the ground (now we have other materials and techniques), but distribute them in groups on paved surfaces and places where the plant is not in direct contact with the soil (greening of terraces). Thus, we can see that Baroque parterres were already quite dynamic structures which could be adopted to a greater extent in today's cities. The idea that plant species were strictly separated in parterres is not quite correct, as it was a common practice to mix different plants and, in some ways, already follow the perma culture as we know it today. Notably, the eighteenth century, which brought an interest in the natural sciences and the development of botany, brought a different perception of plants and their coexistence. Thus, botanical enthusiasts, such as Baron Erberg in Carniola (a part of the present-day Slovenia), began assembling their flower patterns. The Baron notes in his description of the garden from 1822 that red pelargonium and pink evergreen are a good combination[6] even though the difference in height was considerable between the plants at the time (it should be borne in mind that this was a time when pelargonium had only just begun to be cultivated and the plant could then reach 1.5 m in height). Parterres were therefore quite colorful—in terms of color and species. This can also be clearly seen in today's successfully restored gardens, such as the Baroque garden of the Hof manor in Austria. The garden began to emerge after 1725 and was owned by Prince Eugene of Savoy [10]. The idea of renovation was born in 1986, but major works were not completed until 2007 and 2019. Today we can stroll through the representative terraces

---

[5] Parterre (Fr.) is a surface formed with different patterns, usually arranged close to the residence. The parterre should be located near the castle or mansion, as it is best observed from above (from the piano nobile) to make it easier to understand its ornamental pattern. Augustin-Charles d'Aviler, then Dezallier d'Argenville, and others wrote about parterre patterns. We distinguish between several types of parterre: parterre a l'anglaise (rectangular lawn parterre, sometimes lined with flowers); parterre de broderie (embroidery parterre, adorned with a fine interweaving line resembling embroidery, from 1620 to 1720 it was the more common motif of formal gardens; the pattern consists of low sheared bush, flowers, multicolored pebbles, sand, gravel, and the like); parterre de broderie melee de massifs de gazon or Parterre melee (composed of diverse patterns of grass belts); parterre de compartiment (the pattern consists of bands of grass, bush, and flowers, mainly used in the second half of the eighteenth century; it is similar to embroidered the but parterre only maintains symmetry in the longitudinal and transverse axes); parterre de pieces coupees pour des fleurs (the floral parterre was intended primarily for decoration of smaller garden areas); parterre d'eau (incorporating water surfaces); parterre gazon coupe (shaped lawn belts; used after 1720, it replaced the embroidery parterre) [6, 7].

[6] "... Die Streife sind am gefalligsten, wenn niedere Pflanzen dieselben, und zwar ja nicht gedrangt besezen. Die rothe Pelargonien und *Vinca rosea* sind fur jeden Fall allein hinreichend" [16].

Figure 12.
The Baroque gardens of Schloss Hof, Austria.

with the ground floor and water motifs and admire the aforementioned "mixed" ground floor at the greenhouse (e.g., **Figure 12**).

6. High above

When examining historic gardens, one can also come across mentions of roof gardens. Again, we can think back to the Babylonian structures, where they already had troughs filled with soil and an irrigation system and drainage so successfully constructed that the gardens thrived even in high positions on the skeletons of buildings. The old civilizations favored roof gardens (even the ancient Romans). In the Middle Ages, the interest in them somewhat diminished, but one could still find examples of small gardens consisting of flowerpots or similar containers placed on raised fortifications, monasteries, etc. The interest in roof gardens grew again in the eighteenth century, as enthusiasm for the plant world took over all layers of people, and many had only a window shelf on which they could observe the growth of primula, pelargonium, or perhaps hydrangea. Even the kings suffered from such "botanical" fever, among them the Austrian King Franz I. (1768–1835), who was named *Blumenkaiser* or "the flower emperor" because of his passion for plants [17]. The flower emperor arranged a terrace on the roof of his city castle, which housed plant pots, hotbeds, and even a greenhouse. It was a place where the emperor spent a lot of his time [18]. Today's roof gardens are of course technically advanced and much more common in urban structures. One of the reasons for that lies in various studies proving the positive cooling effect of the environment (especially in cities, where solar radiation is absorbed by roads and buildings and this heat accumulates in the building material).

7. Shelter from cold

We have successfully adopted a lot of knowledge from history—bosquetts have been transformed into green walls, fountains into water jets freely arranged on the surfaces of city squares, and various parterre bordures into mixed (permaculture) gardens. Furthermore, facilities for overwintering delicate plants have also been upgraded. Greenhouses flourished in the time of introduction of non-native species into Europe [19]. In the second half of the eighteenth century, the introduction of alien, exotic fruits onto the tables of the nobility brought even greater diversity of such plants. The always fresh and varied fruits and vegetables on the gentleman's table were among the significant qualities of a higher class. They were a kind of status symbol, and many noblemen arranged greenhouses and other winter facilities for their cultivation. In addition to the well-growing figs, lemons, oranges, or pomegranates, melon[7] and pineapple played an important role in the eighteenth century. In Versailles, melons were a popular fruit of the French court (they had to provide 100 melons a day in 1688), but they were also grown in large numbers at many other courts (e.g., at the Prussian court, where even *Melonerie* was held in Sanssouci near Potsdam). Melons retained their popularity, although cultivation of this fruit required a great deal of work [20]. Pineapple was also very popular, as evident from the following description: "The excellency, frangrancy, and flavor of the fruit which this plant produces needs no commendations, as it is well known to excel all the fruits hitherto cultivated; so that it is no wonder every gentleman of taste and fortune is the foundation of this polite article of gardening" [21]. Taste and money were necessary to grow this plant, as it originated from South America and Africa, thus it required a special heated greenhouse. Protective facilities for non-native plants varied and ranged from glass bells that covered individual plants, to low warm beds or large greenhouses where the south glass wall was inclined to maximize the warmth of the sunlight. There were greenhouses that fascinated with new technology in the nineteenth century and, as mentioned greenhouses that were only heated by the sun, a furnace, or a kind of hot water system. The variety of the greenhouses depended also on the plants and of course the (financial) ability of the owner. Today's greenhouses are just as diverse. Sometimes gardeners used (plastic) bottles to protect iindividual plants from frost (which is similar to the practice of protecting the plants from frost with glass bells), while large gardens usually have more sophisticated glasshouses. Maybe modern urban horticulturists could follow the example used in the Rococo garden of the summer palace of Frederick the Great, Sanssouci in Potsdam (built between 1745 and 1747), where a combination of solar heat and partial glazing is used (e.g., **Figure 13**). Namely, the palace stands on the top of a hill that was

Figure 13.
The south-facing garden façade of Sanssouci in Potsdam, Germany.

[7] Until the nineteenth century, melons were considered the perfect fruit to have in the garden. In everyday use, melons were not exactly distinguished—the name "melon" was used for several different species, including melons, pumpkins, cucumbers, zucchini (from the seventeenth century onwards), and other fruits or vegetables [20, 22].

transformed into terraces. Trellised wines from Portugal, Italy, France, and from nearby Neuruppin were planted along the brick walls of the terraces. Between them 168 glazed niches were created in which figs trees grew [9, 22]. An ideal place for a king to relax and forget his worries.

8. Coherent thought

A quick look at the historic gardens and their motives and elements should give a slightly clearer picture of what man has already adapted to his needs in his gardens, and what we can draw from past knowledge. It becomes clear that man has increasingly consciously included nature in his world. Perhaps this was most evident in the 19th century, when Ebenezer Howard (1850–1928) intensely researched the idea of the ideal city and, based on the study of past thinkers (such as the Renaissance architect Filarete), he first created a plan for his ideal city. However, while the past thinkers looked at a city only as a built structure, Sir Howard incorporated the natural and cultural landscape into his city. He introduced his garden city in his publication *To-Morrow: A Peaceful Path to Real Reform* (1898), where one can find a Utopian city in which people live harmoniously together with nature. Thus, at the end of the nineteenth century, Sir Howard understood the city in a wider sense—he saw green-designed areas as essential components of the city. Whether we talk about gardens, parks, river landscapes, or cultural landscapes, it is not as important as the fact that built structures cannot exist without the nature. That was clear to people even before Howard's realization. For example, in the second half of the century, a city park was created in Bremen, Germany. However, the investor was not the ruler or the land, state, or any other corporation. The townspeople themselves set up the park and a society to maintain it [23]. A similar example can be found even earlier in the nineteenth century in the city of Ljubljana (now the capital of Slovenia). At the edge of Ljubljana, the idea of a great tree avenue that would provide a pleasant walking area and a space for amusement and relaxation for the citizens arose under the French government in 1813. In 1814, when the new Austrian government came to power, the project was begun but not completed. Thus, in the same year the initiative to complete the avenue was taken by the citizens, and the so-called Lattermann avenue was completed no later than 1816 [24].

More than a hundred years have passed since Howard's idea (and the realization of his garden city), and more than 200 since the citizens of Ljubljana created their own designed green space on the outskirts of the city. Today, we can praise the utopian idea of the "garden city" and admire the determination of the mentioned townspeople who connected the natural wooded hill and the old city with a designed green structure. In comparison with our surroundings, they had more natural areas at their hand, and they did not need to incorporate as many green areas into the tightly built cities as possible, the need that we have today.

Urban horticulture helps us create and materialize possibilities of incorporating nature into our cities, and we need to seize them fully so that the prediction made by the Swiss curator Klaus Littmann—that in the future we will be observing nature only in isolated spaces, similarly as we today observe animals in a zoo (especially rare or even extinct ones)—does not come true. Between September 8 and October 27, 2019, Littmann carried out a major project of planting an indigenous Carinthian mixed forest (which has almost disappeared from Carithia as it is continuously replaced by much more profitable conifer monocultures) in the Klagenfurt football stadium (e.g., **Figure 14**). This intervention that attracted masses of people is not just

Figure 14.
Littman, For forest (2019), Klagenfurt, Austria.

an art project but also a warning appeal and a warning echo started by artists such as Robert Smithson (his well-known Spiral Jetty created in 1970 is located outside the urban environment, on the North Salt Lake near Rozel Point in Utah), or Christo and Jean-Claude (they wrapped around 178 trees in Basel in 1998) or even Joseph Beuys (he introduced and subsequently implemented the project *7000 Eichen—Stadtverwaldung statt Stadtverwaltung*, at Documenta 7 in 1982; [25, 26]).

9. Conclusion

We have investigated historical gardens and their water motives in all their variety. We have taken a peek at small garden objects—those in the form of nature-like cavities—and others following contemporary architectural styles. We have found the similarities between bosquetts and green walls, and seen that the art of cutting trees and shrubs was an art of itself—which is well known by those who preserve historic (mostly Baroque) gardens [27]. We have outlined the diversity of parterres whose its ornamental lines are filled with plants that were changed according to their blooming periods—gardening that resembles today's pot cultivation (or container gardening). At the same time, we saw that the plants in the bordures of parterres were not monoculture, but contained many different species—parterres were closer to today's understanding of permaculture. We also briefly discussed roof gardens and green houses for more delicate plants. In these examples the enthusiasm for botany was presented—the enthusiasm that was not foreign to many rulers of the late eighteenth and nineteenth centuries. Kings and queens included new plants, their exploration, and the designing of gardens (where they liked to include different rare plants) in their schedules, devoting much of their time and finances to this love of botany. Not only did they build large botanical gardens, greenhouses, and collections (the most famous example is certainly the Kew gardens, royal botanical garden in England), but they also had small gardens just for themselves (sometimes on the roofs of their castles). They studied plants in their botanical cabinets, collected botanical books, and made notes, herbarium, and botanical journeys. This enthusiasm was the result of the development of interest in natural science, which has evolved into disciplines that today provide us with the knowledge of plants to the extent that we can integrate nature into our urban centers.

Furthermore, we have seen that, especially in the nineteenth century, urban residents understood the importance of green spaces in the city, to the extent that they themselves (at their own initiative and at their own expense) set up city parks. Even though the green surroundings of the city were still unspoiled at the time, and the cities were not as big and densely built as they are today, they knew how important it was for a person to have access to the natural environment every day. The emphasis on this importance is attributed to the industrial revolution, which, in addition to technological advancement, brought with it a well-defined working schedule. Leisure days were rare (initially only one day a week was free), and in those days people loved to spend their time in the soothing embrace of nature (far from noisy and dirty machines and enclosed industrial halls).

So, let us look ahead and make sure that nature is preserved for the next generations, and that it retains in its original form and activity. By integrating different ecosystems in our cities, we can enable this. In this way, plants will contribute to the improvement of living conditions (reducing pollution, reducing the impact of global warming, offering space for relief and contemplation, etc.) as well as provide fundamental links for the nature, enabling it to be coherent, improving conditions for its reproduction and transmission of the information it needs for its existence. Furthermore, with a good insight into the past, we can make the greening of our cities easier. Street façades could all be dressed in green—not only as green walls but also as a structure for trees and shrubs and climbers to grow on. When visiting Viertel Zwei in Vienna, we can see a ten storey "vertical green" residential building, where an additional structure for plants was made in front of the actual façade, creating a tangible green space for the residents (e.g., **Figure 15**). Thus, when creating a new neighborhood in the city, we should devote special attention to its "green" part. Trees should be planted along the streets—all the art of shearing trees and shrubs in Renaissance and Baroque could be used on the narrow streets. Cascades and fountains should be part of city squares; they could cross the edges of pools or stairways and provide play space (or even generate electricity). Lakes and ponds could offer more peaceful areas in neighorhoods. In the spaces between larger apartment buildings, gardens could be arranged that would follow the schemes of Baroque gardens and offer beautiful views of the colourful design of parterres when looked at from higher floors of buildings.

Figure 15.
Vertical green building in Viertel Zwei (built 2017), Austria.

In the centre of such garden, a garden architectural structure could be built, in which a grotta would be arranged in the lower part, and on top of it a room for socializing, listening to music, playing cards, or even a greenhouse. Aquaponics could be included in the grotta and greenhouse system. Green roofs and terraces could be used to grow exotic, heat-loving plants. With all this in mind, let us not forget about water instruments (organs) that could be part of such grottas or they could be used to liven up roadside green patches where rain water in roadside channels could be used.

Nature can be introduced into almost every element of a tightly built city. Following the example set by many art projects—be it wrapping trees in decomposable materials, placing tiny sculptures in a narrow street waiting for rain, planting oaks in the city, etc.—through joint collaboration we should make sure that authentic forests will not be only recreated in stadiums.

Author details

Ines Babnik
Department of Art History, Faculty of Arts, University of Ljubljana, Ljubljana, Slovenia

*Address all correspondence to: ines.babnik@ff.uni-lj.si

References

[1] United Nation. Revision of world urbanization prospects. 2018. Available from: https://www.un.org/development/desa/en/news/population/2018-revision-of-world-urbanization-prospects.html

[2] Relf D. Human issues in horticulture. Hort Technology. April/June 1992;**2**(2):985. DOI: 10.21273/HORTSCI.38.5.984

[3] Clifford D. A History of Garden Design. London: Faber And Faber; 1962

[4] Kluckert E. Gartenkunst in Europa von der Antike bis zur Gegenwart. Königswinter: Tandem Verlag GmbH; 2007

[5] Ogrin D. Krajinska arhitektura (učbenik na študiju krajinske arhitekture). Ljubljana: Oddelek za krajinsko arhitekturo, Biotehniška fakulteta; 2010

[6] Symes M. A Glossary of Garden History. Madison: Shire Publications; University of Wisconsin; 1993

[7] Uerscheln G, Kalusok M. Kleines Worterbuch der europaischen Gartenkunst. Stuttgart: Philipp Reclam jun. GmbH&Co; 2001

[8] Cincinnatiparks. Smale riverfront park. 2019. Available from: https://www.cincinnatiparks.com/river-parks/smale-riverfront-park/

[9] Hansmann W. Gartenkunst der Renaissance und des Barock. DuMont: Koln; 1988

[10] Gothein LM. Geschichte der Gartenkunst I/II. Georg Olms Verlag: Hildesheim, New York; 1977

[11] Chateau Versailles. Gardens. 2019. Available from: http://en.chateauversailles.fr/discover/estate/gardens

[12] Viertel Zwei. Viertel Zwei. 2019. Available from: https://www.viertel-zwei.at/

[13] Ditsche A. Hydropneumatische Automatophone im Gartenbau der Neuzeit. In: Butenschön S, editor. Garten-Kultur-Geschichte-Gartenhistorisches Forschungskolloquium. Berlin: Univ.-Verl. der TU, Univ.-Bibl; 2010

[14] Unetič I. O zelenem okrasju Turjaka. In: Preinfalk M, editor. Grad Turjak. Ljubljana: ZRC SAZU; 2019/20 (forthcoming)

[15] Babnik I. In the gardens of enlighteners in Carniola around 1800. In: Bouchard F, Farinelli P. editors. Les régions slovènes entre XVIIIe et XIXe siècles: plurilinguisme et transferts culturels à la frontière entre empire des Habsbourg et Venise. Tours: Université de Tours, 2019

[16] Archive of Slovenia: SI AS 730, fasc. 5, Dominicalia; Ein Beitrag ..., fol. 950

[17] Pfundheller J. Der Blumenkaiser-Oesterreichischen Zeit- und Culturbild. Wien: Manz'sche k. k. Hof- und Universitat-Buchhandlung; 1881.

[18] Lack HW. Florilegium Imperiale, Botanical Illustrations for Francis I. of Austria. Prestel: Munich, Berlin, London, New York; 2006

[19] Heinze, Kristin H. Das "Treibhaus" als Metapher für eine widernatürliche Erziehung im Kontext der sich im18. Jahrhundert herausbildenden Pädagogik als Wissenschaft. In: Eggers M., Rothe M.(ed.). Wissenschaftsgeschichte als Begriffsgeschichte. Terminologische Umbrüche imEntstehungsprozess der modernen Wissenschaften. Bielefeld: transcript, Verlag. 2009. p. 107-131. DOI: 10.14361/9783839411841-004

[20] Wimmer AC, editor. Die Melonen der Monarchen. Berlin: H. Heenemann GmbH; 2003

[21] Giles J. Ananas: Or, a Treatise on the pine-apple. In: Which the Whole Culture, Management and Perfecting This Most Excellent Fruit .... London: S. Blandon. p. 1767

[22] Heilmeyer M, Wimmer CA, Schurig G. Die Melonen der Monarchen. Potsdam: Vacat; 2003

[23] Chevallerie de la H. Tivoli, ljubljanski mestni park. In: Strgar I, editor. Tivoli. Kmečki glas: Ljubljana; 1994

[24] Unetič I. Kultura vrtov. Znanstvena založba Filozofske fakultete: Ljubljana; 2016

[25] Waldner-Petutschnig K. Wie ein Wald ins Stadion kommt. In: For Forest. Klagenfurt: Kleine Zeitung; 2019

[26] Rosenthal M. Understanding Installation Art. Prestel: Munchen, Berlin, London, New York; 2003

[27] Doblhamer R. Gartendenkmal pflegerische und Gartentehnische Überlegungen zum Austausch der Heckenwände im Grossen Partere von Schönbrunn. Österreichische Zeitschrift für Kunst- und Denkmalpflege. 2003; 57(3/4):399-418

Chapter 10

Implication of Urban Agriculture and Vertical Farming for Future Sustainability

Anwesha Chatterjee, Sanjit Debnath and Harshata Pal

Abstract

Urban agriculture (UA) is defined as the production of agricultural goods (crop) and livestock goods within urban areas like cities and towns. In the modern days, the urbanization process has raised a question on the sustainable development and growing of urban population. UA has been claimed to contribute to urban waste recycling, efficient water use and energy conservation, reduction in air pollution and soil erosion, urban beautification, climate change adaptation and resilience, disaster prevention, and ecological and social urban sustainability. Therefore, UA contributes to the sustainability of cities in various ways—socially, economically, and environmentally. An urban farming technology that involves the large-scale agricultural production in the urban surroundings is the vertical farming (VF) or high-rise farming technology. It enables fast growth and production of the crops by maintaining the environmental conditions and nutrient solutions to crop based on hydroponics technology. Vertical farms are able to grow food year-round because they maintain consistent growing conditions regardless of the weather outside and are much less vulnerable to climate changes. This promises a steady flow of products for the consumers and a consistent income for growers. Various advantages of VF over traditional farming, such as reduced farm inputs and crop failures and restored farmland, have enabled scientists to implement VF on a large scale.

Keywords: urban agriculture, vertical farming, hydroponics, aquaponics, aeroponics

1. Urban agriculture

Urban agriculture (UA) can be defined as the growing of plants and rearing of livestock within a city (intra-urban) or on the areas surrounding the cities (peri-urban agriculture), involving input provision and processing of raw materials into edible forms followed by marketing activities [1, 2].

2. Need for urban agriculture/importance of urban agriculture

The proportion of the world's population living in cities is increasing dramatically. It is predicted that by 2030, the worldwide population of urban dwellers will be nearly 5 billion [3], and by 2050 it may reach 9 billion [4]. The increased rate of urbanization has important economic, social, and political implications: A large number of people residing in the cities can approach toward education and employment easily; they can trust the healthcare industry and can see cultural evolution. But this rapid growth of population is often integrated with communal challenges and also climate change: cities may fail to provide the basic facilities resulting in communal riots leading to inferior and undesirable living conditions. Therefore, in order to deal with the challenges of rapid urbanization, urban agriculture is in demand nowadays.

The need or importance of urban agriculture is broadly discussed with the following *advantages* associated with it [5].

2.1 Fighting environmental challenges

Today, cities consume more than two-thirds of the world's energy and are responsible for 70% of global CO_2 emissions. Recently, UA is considered to deal with the difficult situations like climate change as it plays sufficiently in greening the metros and improving the warmer city climate while encouraging the reuse of organic wastes that reduces the urban energy footprint [2].

The World Meteorological Organization (WMO) suggested that more urban farming should take place as a response to climate change and as a way to build more resilient cities.

UA helps cities to improve the urban environment and become more resilient by [2, 6]:

- UA reduces the weakness of specific urban groups and diversifies urban food sources and income opportunities of the urban poor and forms a source of innovation and learning about new strategies/technologies for land- and water-efficient food production.

- UA helps in keeping the open areas covered with greeneries that might reduce the severity of the climatic conditions. UA also makes the microclimate worth living and also forbids the construction of buildings on risky areas, and by this not only flooding, landslides, and other disasters are reduced but also urban biodiversity and living conditions are improved. Such open green spaces also help to control storm water flows by allowing water storage and increased infiltration of excess storm water [7]. In these open green spaces in and around urban areas, food production can be combined with other services to city dwellers, such as agro-tourism or park and landscape maintenance, e.g., "productive parks."

- UA produces fresh green foods that reduces the green-house gas emission and also uses limited energy in the process of getting food from the farm to the plate in industrially developed countries [8].

- Productive reuse of waste water in UA helps to combat the freshwater crisis and also saves rivers, canals, and other water bodies from being polluted by the waste water. On the other hand, waste water as a source of irrigation might decrease the risk of water scarcity [9]. Use of urban waste water as a source of irrigation will help to adapt to risks of drought and flood. Urban waste water can be recycled for irrigation/fertilization of horticultural crops, i.e., floriculture and fruit crops, as well as for irrigation of forest plantations that provide wood for fuel.

2.2 Food security and nutrition

UA contributes to enhance urban food security and nourishment of the poor class. Families that are involved in UA are exposed to better quality and variety of diet. They consume more herbs and greens than the others. Production of food by urban families can supply up to 20–60% of their total food consumption especially in green vegetables, medicinal and aromatic plants, eggs, and milk and meat from small animals. Involvement in UA may also cause better mitigation of diseases as it has better nutritional and medicinal properties in homegrown medicinal plants, it causes more physical exercise, and people do not have to depend on gifts and food aid which may enhance their self-esteem. UA also increases the accessibility of fresh and affordable food for other urban consumers, as most of the food produced by urban farmers is bartered or sold locally. UA also ensures food requirement during natural calamities and wars. In Sierra Leone, the residents devoted themselves in UA in order to meet their daily foods during the civil war that lasted for about 10 years. UA acts as a survival strategy for the refugees and helps them to live in a state of being worthy for honor [1, 2, 6].

2.3 Poverty alleviation

The world's urban population is expected to reach 6–9 billion by 2050. It is estimated that poverty will progress from villages to the metro cities by 2030 as 60% of the Earth's population will reside in the cities. Moreover, in most developing countries, urbanization has led to the growth of slum population which has almost doubled in the past 15 years [3]. Also this rapid urbanization in developing countries created difficulty in making sufficient employment opportunities creating very poor living conditions in the slum areas. The presence of UA can definitely meet the requirement of employment to some extent in the cities of developing countries. The effects of UA on poverty alleviation vary with the type of participants involved, the products produced, and the degree of market orientation, among other things. UA often plays an important role in the survival strategies of the urban dwellers, who might be benefited from UA in various ways: Firstly, when a household produces edible crops, their food expenses are reduced and they can do a huge amount of savings. Moreover, the surplus produce can be sold by them in order to make a profitable business [2, 6].

2.4 Proper land use

In addition to climate change and urbanization, food production is confronted with decrease in productive agricultural land. Large-scale urban food production could provide opportunities and take the pressure off agricultural land. Consequently, researchers and practitioners are aiming to separate arable land from production and produce food on a larger scale in and on buildings in high-density urban areas. Scientists visualized the "edible city" and introduced the concept of continuous productive urban landscape (CPUL), recommending the coherent introduction of interlinked productive landscapes into cities as an essential element of sustainable urban infrastructure. One major challenge of urban food production is land availability and access. Principally, there might be large resources of land that could be made accessible for agricultural purposes, but for densely built-up areas and where availability of space often limits the area of production unit, no-space or low-space technologies provide opportunities for space-confined growing [5, 10, 11].

Besides its so many advantages, there are some *disadvantages* of UA associated with potential health risks [2]:

1. Reuse of contaminated, untreated irrigation water from urban streams gives rise to potential health risks. This can be managed through complementary health risk reduction measures as explained in the 2006 WHO guidelines for safe use of excreta and waste water.

2. Insufficient or improper management of livestock leads to health risks. Proper management of animals, manure, urine, and slaughterhouse procedures will reduce the rate of the associated health risks.

3. Intensive use of fertilizers, pesticides, and fungicides in UA may lead to residues of agrochemicals in crops or in the groundwater. The risk mainly occurs in areas with commercial urban farming. In subsistence and semicommercial urban farming, this risk is limited because the producers rarely apply agrochemicals due to poverty. They use composted organic wastes as they prefer a clean product for self-consumption.

3. Vertical farming: an urban farming technology

With rapid worldwide population growth, there is scarcity of agricultural lands. It increases the demand for both more food and more land to grow food. But some entrepreneurs and farmers are beginning to find a solution to this problem, one of which can be found in the abandoned warehouses in our cities, in new buildings built on environmentally damaged lands, and even in used shipping containers from ocean transports. This solution is called vertical farming, which is an UA technology involving growing crops in controlled indoor environments, with precise light, nutrients, and temperatures.

In vertical farming, growing plants are arranged in layers that may reach several stories high. Although small-scale, residential vertical gardening (including window farms) is under practice for several years, commercial-scale vertical farms have become an important topic of discussion for the past few years in the United States. This new farming technology is growing rapidly, and entrepreneurs in many cities are taking an interest in this innovative farming system [12].

Vertical farming is gaining its importance throughout several urban cities around the world due to the beneficial role it plays in the field of agriculture. Vertical farming can reduce the transportation costs due to its adjacency to the buyer; planned production of herbs and their growing conditions can be enhanced by adjusting the temperature, humidity, lighting conditions, etc. Indoor farming in a controlled environment needs much less amount of water than outdoor farming because it involves recycling of waste water. Because of these features, vertical farming is widely implemented initially in desert and drought-stricken regions, such as some Middle Eastern countries, Africa, Israel, Japan, and the Netherlands [13].

4. Types of vertical farms

4.1 Hydroponics

It is the predominant growing system used in vertical farms, involving growing plants in nutrient solutions that are devoid of soil. The plant roots are submerged in a nutrient solution, which is frequently examined and circulated to ensure that the correct chemical composition is maintained [12].

Urban hydroponics is not a recent invention. The Hanging Gardens of Babylon and the Floating Gardens of the Aztecs were beautifying the cities for quite a long period of time. Also, fruits and vegetables were cultivated in those areas. Nowadays, modern cities use urban hydroponics for physical and psychological relaxation. It is also plays an important role in managing the urban environment. In areas with arid climate, it increases humidity and lowers temperatures. It also captures dust and polluted air by the foliage of the plants. It contributes to the reduction of the overall discharge of CO_2, hence preventing global warming to some extent. Hydroponics gardens are usually constructed vertically because city space is limited. Apart from immediate improvement in the environmental quality, vertical farms on top of traditional buildings serve as large heat sinks that radiate heat and increase ambient air temperature; hydroponic systems thermoregulate buildings by trapping heat in the winter and cooling buildings in the summer. The air quality inside the house can also be improved by growing plants on interior walls. In some modern cities, for example in Bangkok, the concrete roads and railway overpasses are covered with hydroponically grown ornamentals. Also commercial centers are decorated with indoor hydroponics for an improved air quality inside [14].

4.2 Aquaponics

The hydroponic system is taken one step forward by another system called aquaponics which combines plants and fish in the same ecosystem. The nutrient-rich wastes produced by the indoor-grown fish serve as feed source of the plants present in the vertical farm. On the other hand, the plant filters and purifies the waste water which is then recycled into the fish tanks [12].

This combination of systems is cheaper and easier as mineral nutrients are not be purchased and the plants are growing totally organically and moreover no additional expenses are required to clean the fish tanks and there is no scene of pesticides harming the fish. Thus, aquaponics is not only cost-effective but also diseases in the systems can be reduced and a very suitable urban farming technology can be formed. Canadian scientist Savidov explained that possibly the organic components in the system make the trace elements readily available to the plant for proper growth and thus recirculating aquaponic system decreases root diseases in the crop with increased crop yield from aquaponics compared with conventional hydroponics. Also fruits and vegetables grown in aquaponic system qualify for organic product certification very easily since no pesticides and fertilizers are used in this system. Some scientists are planning to construct vertical farms in skyscrapers and have created the name sky farming. Such buildings may also incorporate aquaponics to ensure a good source of fresh fish [14].

4.3 Aeroponics

This innovative indoor growing technique was first developed by the National Aeronautics and Space Administration (NASA). In the 1990s, NASA started finding efficient ways to grow plants in space and coined the term "aeroponics." Aeroponic systems are still in a growing phase in the vertical farming world, however gaining interest gradually. It is an efficient plant-growing system in vertical farms, using up to 90% less water than other efficient hydroponic systems. Plants grown in these aeroponic systems take up more minerals and vitamins, making the plants healthier and more nutritious [12].

In tropical hot and humid climate, it is difficult to grow temperate vegetables like lettuce. Geoff Wilson, an agricultural journalist and Australia's representative of a group of 16 national organizations for an international Green Roofs organization, has reported in an article that a new aeroponic system originated in Singapore can provide a solution to this difficulty. Traditional aeroponic method involved cold nutrient mixture that used to be sprayed onto the plant roots, thereby lowering the temperature causing wilting and ultimately death of the plant. But this type of cooling is expensive, even for rich cities like Singapore. To overcome this limitation, in the year 2004, Gregory Chow, lecturer at the Ngee Ann Polytechnic of Singapore invented the air dynaponics—a much less costly way of maintaining low root-zone temperatures for commercially successful aeroponics. This system gave positive outcomes. Researchers stated that the nutrients infused with oxygen "energized" the entire root system and improved the plant top biomass. Air dynaponics uses the cooling methods of Venturi nozzle effect in an air-powered operation that lowers the temperature of the nutrient mixture and supplies air from the dissolved oxygen. In Singapore, this method is used to produce valuable greens like butterhead lettuce, Batavia lettuce, and Romaine lettuce for moneymaking purposes [14].

4.4 Vertical farming systems can be further classified on the basis of structure that houses the system

4.4.1 Building-based vertical farms

These are the types of vertical farms constructed in abandoned buildings in urban areas. For example, Chicago's "The Plant" vertical farm was constructed in an old pork-packing plant. Vertical farms are also constructed in new buildings. A new multistory vertical farm is built to an existing parking lot structure in downtown Jackson Hole, Wyoming. Here, vegetables are grown throughout the year in the 13,500-square-foot hydroponic greenhouse for sale to restaurants, to local grocery stores, and also directly to consumers [12].

4.4.2 Shipping-container vertical farms

These types of vertical farms are becoming popular day by day. They use 40-foot shipping containers that carry goods around the world and house vertical farms with LED lights, drip irrigation systems, and vertically stacked shelves for growing a variety of plants. It contains computer-controlled growth management systems that allow users to examine all systems from a smartphone or computer. The three leading companies producing shipping-container vertical farms are Freight Farms, CropBox, and Growtainers [12].

5. Advantages of vertical farming

Despommier mentioned a number of environmental and social advantages in his book called "The Vertical Farm: Feeding the World in the 21st Century." The advantages are summarized below [15]:

- Vertical farming ensures production of greens all year round in nontropical countries and is better than normal farming. Despommier stated that 1 acre of vertical farm can produce products almost equal to the amount of products produced by 30 acres of normal farmland on considering the number of crops produced each season.

- Vertical farming involves reduction or abandonment of the use of herbicides and pesticides. In some cases, vertical farming uses ladybugs and other biological controls when required.

- As the crops in a vertical farm are grown under a controlled environment, they are safe from extreme weather conditions such as droughts, hail, and floods.

- Hydroponic growing techniques help in water conservation by using about 70% less water than normal agriculture.

- Indoor farming reduces or eliminates the use of tractors and other large farm equipment that are commonly used on outdoor farms, thus reducing the burning of fossil fuel. According to Despommier, large-scale vertical farming could result in a significant reduction in air pollution and in CO_2 emissions.

- Vertical farming is people friendly. Some hazards that can be avoided in vertical farming are accidents while operating heavy farming equipment and exposure to harmful chemicals.

6. Disadvantages of vertical farming

Apart from so many advantages, there are many critics of the vertical farming described by the scientists. They claimed that there are a number of problems in vertical farming. The challenges to vertical farming may be summarized as follows [13]:

- Start-up costs are high in order to purchase land in central business districts.

- The number of crops grown is sometimes less than rural farming.

- Production volumes are also not as large as conventional farming and scaling-up may add cost and complexity.

- Raising investment capitals and training a skilled workforce are also challenges in vertical farming.

7. World-wide implementation of urban agriculture/vertical farming

Scientists explored the motivations for urban gardening in Germany by screening 657 urban gardening project websites and characterized the types of gardeners, cultivation methods, and consumer behavior. The study also highlighted the "terrabioponic smart-garden system" where the plants grow in natural soil and in organic nutrient solution, which may facilitate social transition toward bio economy [16]. Also scientists from the United Kingdom reported that vertical farming system has increased the yield of lettuce per unit area as compared to traditional horizontal hydroponics [17]. Agriculture and food production activities in the cities of Mexico can contribute in reducing carbon footprint by creating green environment and better land use [18].

Case study 1: The world's largest indoor vertical farm, AeroFarms, is located in Newark, New Jersey, which grows more than 2 million pounds of greens per year without sunlight, soil or pesticides. Instead of using a huge quantity of water to grow plants, AeroFarms system sprays nutrient-rich mist to the plants. Seeds are sown, germinated, and grown on reusable sheets of cloth and are stretched out over trays stacked vertically. LED lights are used instead of the sun, and the exposure is controlled depending upon the maturity of the plant [12].

Case study 2: Rob Laing founded Farm.One in the year 2016 in order to grow rare and hard-to-find produce to the chefs and restaurants in the middle of New York City. The first farm was set up at the Institute of Culinary Education (ICE) in downtown Manhattan, and the second farm is in Tribeca. It uses hydroponics and LED lights and aims to grow rare produce every year. The company supplies rare herbs, edible flowers and microgreens to some of the best chefs in New York [19].

Case study 3: One of the world's first commercial vertical farms named Sky Greens was built in Singapore. This vertical farm produces one ton of vegetables every other day. Large varieties of tropical vegetables like Chinese cabbage, spinach, lettuce, xia bai cai, bayam, kang kong, cai xin, gai lan, and nai bai are grown. Sky Greens uses a hydraulic system called "A Go-Gro," which consists of 6-m-tall hydraulic water-driven A-shaped towers. Each tower contains 22–26 tiers of growing troughs, and is spun around the aluminum frame at a speed of 1 mm/sec for a steady radiation of sunlight, proper air flow, and irrigation for all the edibles growing in the tower. The rotation system is powered by a unique gravity-aided water-pulley system that uses only 1 L of water per 16-hour cycle, which is collected in a rainwater-fed reservoir. The water used in powering the frames is recycled and filtered before returning to the plants. The organic wastes produced on the farm are composted and reused [19].

8. Concept of urban agriculture/vertical farming in India

India is one of the largest producers of fruits, vegetables, and many other agricultural products. In India, vertical farming has been introduced in recent times. Experts from Indian Council of Agricultural Research (ICAR) are working on the concept of "vertical farming" which can be implemented in metros like New Delhi, Mumbai, Kolkata, and Chennai [20].

8.1 Current scenario of urban agriculture/vertical farming in India

Scientists at Bidhan Chandra Krishi Viswavidyalaya in Nadia, West Bengal, had initial success on growing brinjal and tomato hydroponically on a small scale. Punjab also has succeeded in producing potato tubers through vertical farming [20].

In cities like Cuttack and Nagpur, the slum dwellers performed organic farming on terrace and plots and sold the surplus products to the local markets. In Delhi, on the fertile banks of Yamuna River, extensive farming is going on in spite of the fact that farmers do not have any legal sanction to do farming there. In Hyderabad, farmers living along the banks of Musi River use water from the river for urban farming and contributed rice and vegetables to the market [21].

In the urban areas of Tripura, to help the youth for income generation, a prototype model on "vertical farming system" was developed. The area of the structure was about 630 sq. ft. with two floors and two galleries. The ground floor contained two cages (50 sq. ft. each) at both corners that accommodated 100 layer chicks. The central space (140 sq. ft) housed 200 bird broiler/layer chicks per batch. Eight goats were kept on the first floor (140 sq. ft.) area. There were also 12 rabbits kept in hanging cages (4 sq. ft. each). Proper drainage facility was provided to collect wastes with storage facility where it was decomposed and used for manuring the pots. Three *Azolla* tanks were constructed above the rabbit cages which were the source of nutrient to the goat as well as the birds. Ten benches (30 cm each) were kept on both sides of the structure which contained 160 pots for growing small fodder, vegetables, and spices. A water tank of 400 L capacity was also provided on top of the structure for storing water for animals and poultry and also providing irrigation to each pot through drip irrigation system [22].

Ideafarms is an Indian design-in-tech company which produces vertical farm products, and the produce is of high quality and organic and the supply is huge. A Bengaluru-based start-up company named Greenopia is selling kits with self-watering pots, enriched soil, and better quality seeds. A Mumbai-based start-up firm U-Farm Technologies is using hydroponic gardening technique to build vertical farm for an individual apartment or for a supermarket [20].

Vertical farming is definitely a solution to critical problems in Indian farming like lack of supply of farm produce, overuse of pesticides and fertilizers, and even unemployment. But there are some challenges: The initial huge cost of infrastructure for implementing vertical farming in India is difficult. Vertical farming in India has to face other challenges like public awareness, technical knowledge, and high cost of managing and maintaining the vertical farm systems [20].

9. Conclusion

Urban farming, both vertical farming or farming on vacant open spaces, can be a favorable way for ensuring food security in India and around the world in the future. Although countries like Europe, the USA, and Singapore have already implemented vertical farming and are dealing with big projects for future concerns, India still has a long way to go as it is restricted to only few self-interest-driven projects. Institutional support, awareness of the benefits associated with urban agriculture, and financial and technological support from the government can only attract the city dwellers and help them to move forward with the concept of urban agriculture in India. Progressive growth of urban agriculture can act as an urban regeneration tool for the cities by providing social interaction and increasing job opportunities and environmental benefits to the urban areas across the globe. Thus, to combat the challenges associated with rapid increase in population, the topic of "urban agriculture" is being closely monitored by scientists, city planners, and the sustainable agricultural community for a better future.

Author details

Anwesha Chatterjee[1], Sanjit Debnath[2] and Harshata Pal[1]*

1 Amity Institute of Biotechnology, Amity University, Kolkata, West Bengal, India

2 Bidhan Chandra Krishi Viswavidyalaya, Nadia, West Bengal, India

*Address all correspondence to: hpal@kol.amity.edu

References

[1] Smit J, Nasr J, Ratta A. Urban agriculture: Food, jobs and sustainable cities. Urban Agriculture Yesterday and Today. 2001. Chapter 2

[2] Zeeuw HD. Cities, climate change and urban agriculture. Urban Agriculture magazine, number 25. September 2011;**25**:39-42

[3] United Nations Population Fund. State of world population 2007, unleashing the potential of urban growth. 2007

[4] United Nations. World population to 2300. 2004

[5] Specht K, Rosemarie S, Hartmann I, Freisinger UB, Sawicka M, Werner A, et al. Urban agriculture of the future: An overview of sustainability aspects of food production in and on buildings. Agriculture and Human Values. 2014;**31**:33-51. DOI: 10.1007/s10460-013-9448-4

[6] Zeeuw HD, Veenhuizen RV, Dubbeling M. The role of urban agriculture in building resilient cities in developing countries. Journal of Agricultural Science. 2011;**149**:153-163. DOI: 10.1017/S0021859610001279

[7] Dubbeling M, Campbell MC, Hoekstra F, René van V. Building resilient cities. Urban Agriculture magazine; June 2009;**22**

[8] Heinberg R, Bomford M. The food & farming transition: Toward a post carbon food system. Post Carbon Institute, Spring. 2009

[9] Buechler S, Mekala GD, Keraita B. Wastewater use for urban and peri-urban agriculture. Cities Farming for the Future: Urban Agriculture for Green and Productive Cities. Netherlands, Canada, Philippines: RUAF Foundation, IDRC, IIRR; 2006

[10] Bohn K, Viljoen A. The edible city: Envisioning the continuous productive urban landscape (CPUL). The CPUL City Concept. Field: A free Journal for Architecture. 2011;**4**(1):149-161

[11] Dubbeling M. Integrating urban agriculture in the urban landscape. Urban Agriculture Magazine. September 2011;**25**

[12] Birkby J. Vertical Farming; 2016. pp. 1-12

[13] Benke K, Tomkins B. Future food-production systems: Vertical farming and controlled-environment agriculture. Sustainability: Science, Practice and Policy. 2017;**13**(1):13-26. DOI: 10.1080/15487733.2017.1394054

[14] Schnitzler WH. Urban hydroponics for green and clean cities and for food security. Acta Hort, ISHS 2013. 2013;**2050**:13-26

[15] Despommier D. The vertical farm: Controlled environment agriculture carried out in tall buildings would create greater food safety and security for large urban populations. Journal für Verbraucherschutz und Lebensmittelsicherheit. 2011;**6**:233-236. DOI: 10.1007/s00003-010-0654-3

[16] Winkler B, Maier A, Lewandowski I. Urban gardening in Germany: Cultivating a sustainable lifestyle for the societal transition to a bioeconomy. 2019;**11**:801. DOI: 10.3390/su11030801

[17] Touliatos D, Dodd IC, Mcainsh M. Vertical farming increases lettuce yield per unit area compared to conventional horizontal hydroponics. Association of Applied Biologists; June 2016. DOI: 10.1002/fes3.83

[18] Anda JD, Shear H. Potential of vertical hydroponic agriculture in Mexico. Sustainability. 2017;**9**:140. DOI: 10.3390/su9010140

[19] Loman LJ. Vertical farming: Can it change the global food production landscape? Nuffield International Project No. 1601. 2018

[20] Sonawane MS. Status of vertical farming in India. International Journal of Applied Science and Technology. 2018;**9**(4):122-125. DOI: 10.15515/iaast.0976-4828.9.4.122125

[21] Ali F, Srivastava C. Futuristic urbanism—An overview of vertical farming and urban agriculture for future cities in India. International Journal of Advanced Research in Science, Engineering and Technology. 2017;**4**(4):3767-3775

[22] Singh AK, Das D. Integrated vertical farming system an innovative way of efficient utilization of small land and farm resources in urban areas. Indian Farming. 2018;**68**(06):23-24